Indochina Memoir

Arizona State University
Program for Southeast Asian Studies
Monograph Series Press
Also Available

Southeast Asia – General Interest

Hill Farms and Padi Fields: Life in Mainland Southeast Asia
Robbins Burling

Modern Southeast Asian Literature in Translation: A Resource for Teaching [Syllabi for Teaching Modern Vietnamese, Indonesian, Thai, and Philippine Literature in Translation]
Grant Olson, editor

Southeast Asian Urban Environments: Structured and Spontaneous
Carla Chifos and Ruth Yabes, editors

Burma/Myanmar

An American in Burma, 1930 to 1942
Muriel Sue DeGaa Upfill

Cushing's Shan – English Dictionary: A Phonetic Version
Thomas John Hudak, editor

Cambodia

Prolific Survivors: Population Change in Cambodia, 1975–1993
Jacqueline Desbarats

China

The Jingpo: Kachin of the Yunnan Plateau
Zhusheng Wang

Kam – English Dictionary
Thomas John Hudak, editor

Indonesia

Gamelan Stories: Tantrism, Islam, and Aesthetics in Central Java
Judith Becker

The Javanese in Suriname: Ethnicity in an Ethnically Plural Society
Parsudi Suparlan

The Pesantren Tradition: The Role of the Kyai in the Maintenance of Traditional Islam in Java
Zamakhsyari Dhofier

Purifying the Faith: The Muhammadijah Movement in Indonesian Islam
James Peacock

Structuralism's Transformations: Order and Revision in Indonesian and Malaysian Societies — Papers Written in Honor of Clark E. Cunningham
Lorraine V. Aragon and Susan D. Russell, editors

Toward a New Paradigm: Recent Developments in Indonesian Islamic Thought
Mark Woodward, editor

The Usage of the Traditions of the Prophet in Contemporary Indonesia
Howard Federspiel

Laos

New Laos, New Challenges
Jacqueline Butler-Diaz, editor

Taking Refuge: Lao Buddhists in North America
Penny Van Esterik

Malaysia

see **Indonesia:** Structuralism's Transformations — *Aragon and Russell*

Philippines

Patterns of Power and Politics in the Philippines: Implications for Development
James Eder and Robert Youngblood, editors

Thailand

Of Mechanical Swans and Rain-Producing Elephants: A Collection of Thai and Lanna Thai Tales and Sermons
Konrad Kingshill, compiler/editor

Thai Tellings of Phra Malai: Texts and Rituals Concerning a Popular Buddhist Saint
Bonnie Pacala Brereton

Vietnam

Minot's White Tai Dictionary: A Phonetic Version
Thomas John Hudak, editor

Southeast Asian Linguistics Society Series

Papers from the First Annual Meeting of the Southeast Asian Linguistics Society, 1991
Martha Ratliff and Eric Schiller, editors

Papers from the Second Annual Meeting of the Southeast Asian Linguistics Society, 1992
Karen Adams and Thomas John Hudak, editors

Papers from the Third Annual Meeting of the Southeast Asian Linguistics Society, 1993
Mark Alves, editor

Papers from the Fourth Annual Meeting of the Southeast Asian Linguistics Society, 1994
Udom Warotamasikkhadit and Thanyarat Panakul, editors

Papers from the Fifth Annual Meeting of the Southeast Asian Linguistics Society, 1995
Shobhana L. Chelliah and Willem J. de Reuse, editors

Papers from the Ninth Annual Meeting of the Southeast Asian Linguistics Society, 1999
Graham W. Thurgood, editor

For further information, check out our web site
http://www.asu.edu/clas/asian/pseas.html

Indochina Memoir

Rubber, Politics, and War in Vietnam and Cambodia 1955–1972

BY

Michel M. Michon

Arizona State University
Program for Southeast Asian Studies
Monograph Series Press

Program for Southeast Asian Studies Monograph Series
Arizona State University, Tempe, Arizona 85287-3502

Printed in the United States of America

This publication was funded in part by the
U.S. Department of Education.

∞ The paper used in this publication meets the minimum requirements of the American National Standard for Information Sciences—Permanence of Paper for Printed Library Materials, ANSI Z39.48-1984.

This book is one of a series on Southeast Asia published by the Program for Southeast Asian Studies, Arizona State University, Tempe, Arizona. For more information, see our web site: http://www.asu.edu/clas/asian/pseas.html or contact us via e-mail: pseas@asu.edu.

Photo Credits: All photos are by Photo Nadal, Saigon, and are used by permission of André Brives.

Cover Photo: Southeast Asia Rubber Plantation Complex.

Library of Congress Card Number: 00-108319
ISBN 1-881044-24-6

I dedicate this book

to those Americans

who gave their lives

in Vietnam,

and also to all those

who survived the war

but who were wounded

in body, mind, or spirit.

Contents

Page

Illustrations xi

Introduction xiii

Foreword xix

Acknowledgments xxiii

Chapter

1 South Vietnam: The Quiet Period 1

2 I Am Abducted by the Viet Cong 9

3 More Viet Cong Pressure:
The People's Courts 21

4 The Abduction of Jean Pernet 31

5 Rubber, Robbery, and the Viet Cong 43

6 The Slaughter of the South Vietnamese
Military Post at Long Thanh 51

7 Marriage and Reassignment to Cambodia 61

Chapter

8 The Communists In Cambodia 67

9 Daily Life in the Communist Zone 75

10 The Arrest of Por Try 83

11 Air Raids 91

12 Escape 99

Epilogue 111

Appendix One:
How Rubber Is Cultivated 115

Appendix Two:
Geography of Vietnam and Cambodia 133

Illustrations

Map

Rubber Plantations and Capital Cities xxiv

Plantation Photos

Southeast Asia Rubber Plantation Complex . . . . Cover

Rubber Tree Extensions 129

Grafted Rubber Tree 130

Seedling Rubber Tree 130

A Rubber Plantation Factory 131

Worker's Houses 132

Foreman's House 132

Introduction

Michel Michon first encountered Vietnam in 1951 as a young French soldier in what he calls the French-Communist War. He learned Vietnamese, commanded Vietnamese soldiers loyal to France, and was on hand for France's fateful last stand against the Vietminh at Dien Bien Phu.

Michon subsequently returned to Vietnam as a planter and for some fifteen years managed French-owned rubber plantations in South Vietnam and Cambodia. In both countries, he enjoyed long years of relative peace (from 1955 to 1960 in Vietnam, and from 1964 to 1969 in Cambodia) in which he became happily attuned to the life and ways of these adopted lands. In this memoir, he alludes briefly to these quiet, productive years but he concentrates on the troubled ones that followed—years in which the wheels of revolution yielded victory to communist movements in both countries.

Michon viewed this transition from the vantage point of the specialized agribusiness enterprises of which he was part. Rubber plantations were huge rural communities comprising thousands of workers and millions of rubber trees laid out in perfect rows and grids across ten to fifteen thousand acres. In one respect, these communities were discrete and self-contained. Their villages were relatively new, having been constructed by the rubber company for its workers and containing shops, schools, and temples, as well as churches and infirmaries, of their own. Supplies such as ammonia and formic acid flowed into them from the outside along with rubber processing technology and, indeed, the management itself—hence Mr. Michon. The latex that flowed out of them went to industrial consumers abroad.

In another respect, however, the rubber plantations were an integral part of the local society. Plantation wages fueled the local economy. And plantation villages housed not only workers but also their extended families, who had all sorts of links to the wider society. Moreover, they were not geographically isolated. In both South Vietnam and Cambodia, Michon's plantations were located in areas that were hotly contested by the rising communists, on the one hand, and their U.S.-backed non-communist adversaries, on the other. (Following the usage of the civil war in China, Michon calls the latter Nationalists in both countries.) In this memoir, we observe this power struggle through the eyes of a partisan but eminently practical businessman who strives, amidst it all, to keep the latex flowing.

Michel Michon was certainly predisposed to see Vietnamese and Cambodian communists as the enemy from his earliest soldiering days in Indochina. Nothing in his later encounters with them changed his mind. Although he found some communist partisans more reasonable than others, in this memoir they appear consistently as arrogant, brutal, and tiresome. Michon especially hated the endless, repetitive propaganda sessions of the National Liberation Front, or the Viet Cong (VC), as he always calls them. Later, in Cambodia, he preferred to deal with North Vietnamese soldiers rather than with the Khmer Rouge, who seemed to lack the Vietnamese ability to be calculating and were therefore more brutally capricious.

Michon gives the Viet Cong credit for their astute use of psychological warfare, particularly the theater of terror, to which he was more than once an eye-witness. In chapter three, for example, a People's Court that Michon and his villagers were forced to witness condemns and summarily beheads a local man because his sister has married an enemy soldier. The Saigon government and its American allies were no match for this, he says. Indeed, Michon observes Saigon's soldiers to be "stupid, infantile, and . . . unbelievably inept" (57–58) and offers several anecdotes to make his point. In one, forty government soldiers occupy and fortify with sandbags a small depot in the plantation's central village—to serve as a "presence" in a territory already thoroughly penetrated by the Liberation Front. The poorly supervised young men soon lapse into careless habits and go about unarmed, playing their guitars and wooing the village girls. The Viet Cong make quick work of them. After a surprise attack, they leave the naked bodies of their naïve adversaries lying dead in the road alongside their guitars.

Michon rued the communist successes but also adjusted to them. Stopping communism was not his job—to protect itself, his company was officially neutral. So, taking risks, he approached the ambient communists as a vexing and dangerous but not altogether unmanageable variable in the production of rubber. During certain periods, he says, he was "in almost daily dialogue with them" (6). This went on until March 1970, when the Khmer Rouge occupied the area around Chamcar Loeu Plantation in Cambodia and created havoc (exacerbated by American bombing) that was too precarious even for Michon. In those days, he says, the Khmer Rouge "would come into my office and shoot hails of bullets into the walls, requiring impossible things such as several tons of rice, barrels of gasoline, extra food, involuntary 'volunteers,' and more" (100). Reluctantly, several months later, Michon slipped away and returned with his family to France.

Michon's memoirs remind us of an important but not often remarked upon aspect of war, and that is that many quotidian aspects of life simply go on in the midst of it, including, in this instance, the operations of huge, labor intensive, rubber plantations. So even as the currents of revolution sent shock wave after shock wave through Michon's plantations, their small army of workers still managed to clear fresh patches of forest for new trees; still built levees, collected, germinated, and planted hevea seeds, and nurtured the seedlings to maturity; still managed to tap millions of mature trees daily and to collect the raw latex from cups on each tree; and still managed to coagulate the latex and press it into sheets and to smoke it and dry it and grade it and sort it and finally to package it for export. At no time, evidently, even when relations between Michon and his adversaries were at their most tense did the rubber from his plantations stop flowing. Even in the most dangerous times (mid-to-late 1970 and early 1971), he tells us, an underground commerce in imported chemicals, gasoline, paraffin, and great volumes of cash (to pay workers) kept the wheels of rubber production turning at the beleaguered Chamcar Loeu plantation. This was no small accomplishment.

There are many vivid images in this memoir. I think of the Vietnamese men accused of stealing raw rubber whose fingers the Viet Cong abruptly hack off in an act of exemplary justice; of Michon's gunless in-house Nung militia armed only with bicycle chains; of Khmer Rouge and North Vietnamese soldiers, each side clearly identifiable by their uniforms, marching in parallel lines toward Michon's Cambodian plantation on opposite sides of the road (a vivid harbinger of things to come); and I think of Michon's own harrowing exploits as a reluctant wartime go-between on behalf of his colleagues and workers.

Yet one image floats above all of these. This is the image of Michel Michon himself, the white planter in his white planter's clothing standing out amidst a sea of black- and brown-clad Vietnamese and Cambodian workers and villagers. It is a surprising image because it seems to come from an altogether different era. Yet, Mr. Michon's memoir reminds us that, in fact, many concrete vestiges of French Indochina survived well into the postcolonial period. In keeping with this reality, Michon acted both as an agent of western enterprise and as the guardian and benefactor of the plantation communities under his charge. He accepted the responsibilities and risks of both roles and, in doing so, repeatedly exposed himself to grave danger.

Michel Michon observes the events and circumstances of his time in Vietnam and Cambodia from his own unique perspective as a private-sector agribusiness manager and as a Frenchman. He is a partisan observer. But he is also an initiated observer, one who speaks with an authentic and credible voice about a time and place we are still grappling to understand. There is still much to learn about this turbulent, formative period in the history of modern Southeast Asia.

Mr. Michon's memoir is one of a series of first-person accounts published by our monograph series in which participants in the social dramas of modern Southeast Asian history speak from their own memories and in their own voices. We think it is important that these voices and memories, in all of their diversity, be preserved for future generations. Arizona State University is delighted to bring Michel Michon's voice, and his story, into the discussion.

James R. Rush, Director
Program for Southeast Asian Studies
Arizona State University

Foreword

By the time the American people were awakened to Vietnam in 1964, after the Bay of Tonkin incident, the French had maintained an interest there for over one hundred years. One of the prime riches that Indochina offered the world, via French technology, was natural rubber. The heveas, or rubber trees, were in fact brought to Vietnam by the French from South America. The marriage of French industry to Vietnamese labor proved to have an enduring honeymoon that was benevolent to both sides for decades—a situation that communist propaganda conveniently ignored.

From 1951 to 1954, Michel Michon served in the Nam-Dinh province of North Vietnam (Tonkin) as a platoon leader of Vietnamese partisan troops. As one Frenchman in charge of forty Vietnamese moving through villages and rice paddies, his daily existence was one of total immersion in the Vietnamese culture. During his military experience, Michel developed a strong empathy for the non-communist Vietnamese as well as a stronger contempt for the communist Vietminh. In his contact with the Vietnamese, he witnessed the power of the psychological strategies used by the communists.

His contempt of communist ideology was peaked by a Vietminh sniper's bullet that struck his comrade Albert Dorin in the head. He fell dead at Michel's side and a firefight commenced, resulting in several dead Vietminh (including the sniper) as well as a bounty of several badly needed rifles. That day, Michel's survival depended on a mere few degrees of windage, but his losses were nonetheless severe. Perhaps such losses as these provided him with his sense of commitment to Vietnam because, as his tour of duty came to an end, he experienced what he calls the "Asiatic bug"—a kind of fervent craving for the lush, prosperous jungles of Vietnam.

In 1955, after the war's end, his knowledge of the people and their language enabled him to land a job with a rubber company. Beginning as an assistant manager, he later became manager of a rubber plantation at Long Thanh. At the outset, he was unaware that he was taking a second step along a course that would lead him from North Vietnam (1951–1954), to South Vietnam (1955–1964), and finally to Cambodia (1964–1972). For over twenty years, Michel faced the wave of communism that swept across Indochina. His witness to the ruthless indoctrination used by the communists versus the bureaucratic militaries of France, the United States, and South Vietnam attests to the efficacy of psychological over technological warfare. The United States inherited the already seasoned, successful enemy when it entered the fight against communism in Vietnam. In its naiveté, America ignored the lessons taught the French in Indochina and plunged its soldiers into an ideological red tidal wave. As that wave crashed across South Vietnam and Cambodia, Michel Michon rode on the flotsam of the war between an inept super power and an all too crafty communist juggernaut.

Opinions about the Vietnam debacle still vary widely in America today. However, through Michel's eyes, everyone can see that Vietnam was clearly a cause worth fighting for, even though the treachery, commitment, and frugality of the communists proved, in the long run, to be weapons superior to America's bombs and tanks. After all, the Viet Cong and the North Vietnamese Army created a two-headed serpent, and America lacked the psychological edge that the Viet Cong provided for the North Vietnamese. This psychological advantage was maintained through a mock jurisprudence of the most savage kind that Michel, as a local resident, was forced to witness—the so-called due process of the "People's Courts." These courts proved to be much more effective than B-52s or Cobra helicopters, and Michel makes it clear why.

Of course, his investment in Indochina—his years of work, his meeting and marrying Monique, and the raising of their children, Jean-Luc and Dominique, there—gave Michel even more to lose when Saigon and Phnom Penh fell in 1975. Monique, who is Eurasian, had relatives on both sides of the war and, of course, Michel's family has lost contact with those friends and family members they were forced to leave behind.

At the time this is written, Michel is retired and studying at Arizona State University in Tempe. Seeing himself as an old friend of the American people, he is disturbed by the number of unanswered questions they have about the value of this country's fight in Vietnam. In an effort to console us, he shares with us accounts of his own and others' struggles and losses that underscore the worthiness of our involvement. His is a clear testimony that America's effort to protect Vietnam was a laudable if doomed fight.

John Miller, Instructor
English Language Institute
University of Southern Mississippi

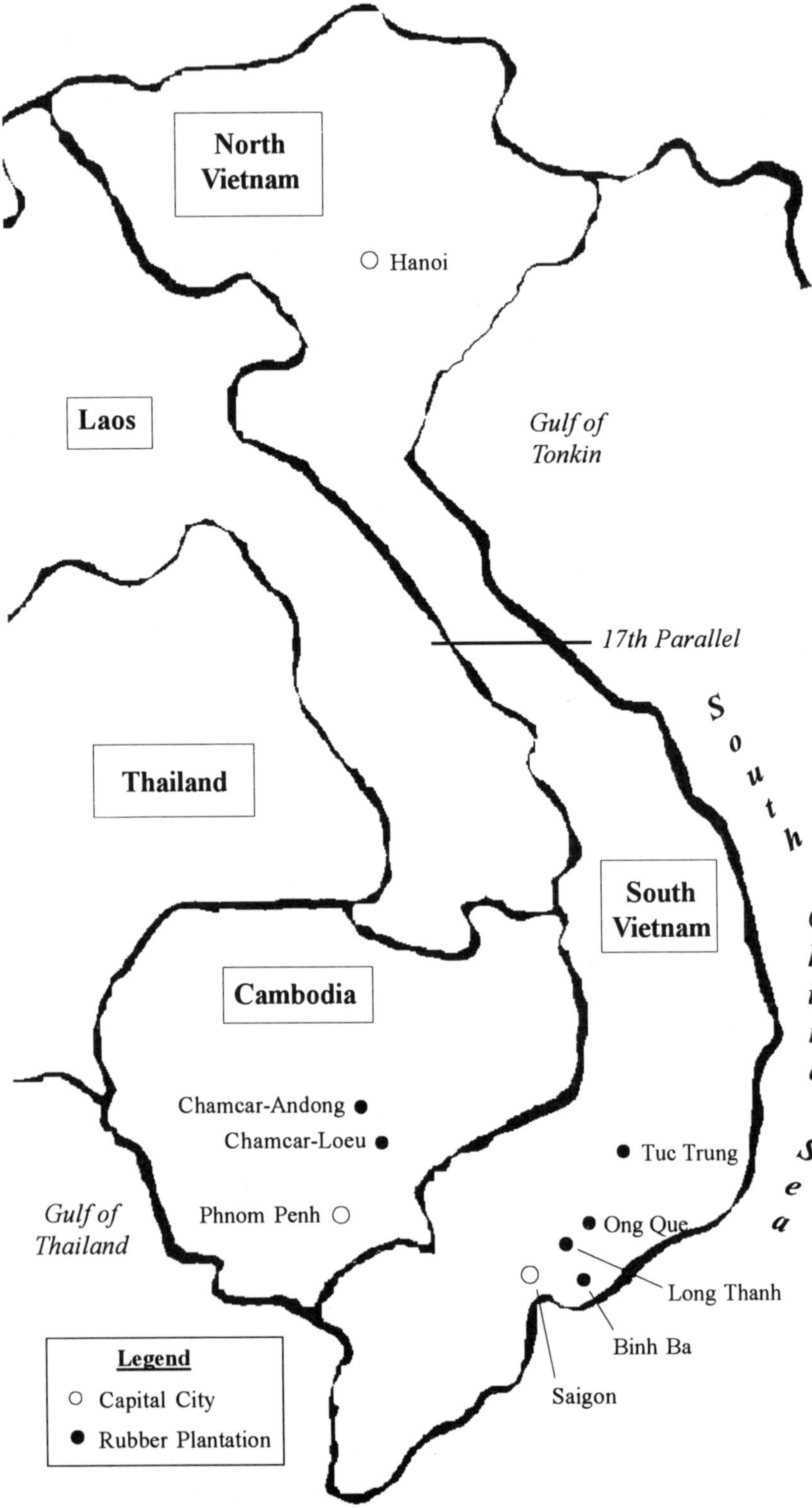

Map: Rubber Plantations and Capital Cities

Acknowledgments

The author is indebted to several people who have helped with their encouragement, positive attitudes, and advice. Recently, I have benefited from the advice of several people at Arizona State University, among whom are the following: James Rush, Dennis Oliver, Mark Rentz, and Nancy Mallette. I especially want to thank my fellow ASU student and Frenchman, Christophe Robert, for contributing his bilingual expertise.

Deborah Boykin, with the Mississippi Arts Commission, Barry Hannah, at the University of Mississippi, and John Guice, at the University of Southern Mississippi, have made unique contributions for which I am grateful.

Glenda Cowart and Yvonne Ross, at the English Language Institute at the University of Southern Mississippi, are thanked for their final touches, as well as their friend and mine, John Miller, who insisted that I share this story with you.

I do not forget Jeanne Taysom, from Tempe, Arizona, without whose help I would not have been able to go through this piece of work.

Finally, I want to thank André Brives who provided from his personal archives the photos of Southeast Asia rubber plantations used in this book. These photos, originally taken by Photo Nadal, Saigon, were used in plantation newsletters produced in the region.

Michel M. Michon
Mesa, Arizona

One

South Vietnam: The Quiet Period

After returning to France in 1954, after more than twenty-seven months (1951–1953) stationed in Tonkin, North Vietnam, with the French Army, I was sufficiently bitten by the "Asiatic bug" to feel like trying my luck in the Far East in a private enterprise. The army could not offer me any opportunity but routine, which I did not want. In fact, I needed to be quite free and have enough initiative in my job in order to feel really responsible about something big and built to last. I did not know if there were many people who had served in a war in a foreign country and still wanted to stay there to work and live. From time to time I wondered if I was not a little crazy. However, the Vietnamese climate was attractive; the people were kind, warm, and tactful; the women were lovely; and I, at twenty-five, was free to do as I liked. Needing to experience something new, I decided to return to Vietnam.

I was lucky enough to meet a childhood friend of mine whose neighbor, at that time on vacation in France, was working for a big rubber plantation company. I met him by appointment with my friend's help and, after a while, I was able to meet the president of the company, who was an approachable aristocrat living in Paris. At this time, the French population in Vietnam was sparse because of the fall of Dien Bien Phu. Also, the Geneva Conference of 1954 had deeply frightened the French people living over there; as a result, most French citizens had returned to France, fearing for their lives and possessions. Most Frenchmen, living in France, were not concerned or interested in Vietnam's or Cambodia's having suffered extensively during the war. My stay in North Vietnam probably influenced the president of this company to hire me because I was fluent in Vietnamese, having been posted to a Vietnamese partisan unit in Tonkin during the war.

In 1955, I found the political climate in Saigon to be rough. Political intrigues between France, the U.S., and South Vietnam; the pending leadership of Ngo Dinh Diem; the guerrilla warfare between South Vietnamese Nationalist forces and three powerful religious sects—the Binh Xuyen, the Hoa-Hao, and the Caodaists; and the massive influx of some 800,000 refugees who had fled Norther Vietnam's communist takeover all contributed to the chaos in Saigon. By mid-1956, Vietnam was calmer with Diem, with U.S. assistance, taking charge in the South, and Ho Chi Minh taking control in the North.

As soon as I arrived at the station where I was to work (Binh Ba in Baria province), I had to begin learning my new job as plantation assistant. After waking up at 4:00 A.M. every day, weekends included, I often worked fourteen to fifteen hours per day without any help. My director, a snobbish, aristocratic old man, let me understand that he did not like young men, much less a former military man like me who had never seen what natural rubber was. So I had to manage the situation and I decided (not having any other choice) to get closer to my Vietnamese workers and their families. They lived in three villages, with about five or six thousand people in each one (700–800 workers with their relatives), and I undertook to visit and talk to everybody in order to have a better knowledge and understanding between each of them and myself. I wanted them to know who I was, what I had done in previous years, how well I could speak their language, and why I was there. I did not want to know what had happened in the past to those on the plantation. My only goal was to produce as much rubber as possible, at the best price, and in such a social climate that every worker, worker's family, the company, and I would all benefit. This practice, which took two or three hours per day out of my personal rest time, was very effective both for my developing fluency in Vietnamese and for my working relationship with the workers. (See Appendix A for an overview of a working rubber-producing plantation.)

I became so successful that, after about a year, no one else at Binh Ba wanted to face the director. I was somewhat embarrassed by acting as *tacit* spokesman because the situation between the director and myself had become very strained. Finally, the general director, who wished for peace but liked efficiency, separated us by appointing me to Tuc Trung, a smaller plantation. Because of its size, there was no director at Tuc Trung and, with my immediate superior fifty miles from me, I would be the lone manager. Therefore, although given the title "chief assistant," I was, in actual fact, functioning as a director. There, where I was no longer a greenhorn, I had the great satisfaction of welcoming several dozen of my former workers and their families, who preferred to join me rather than stay in Binh Ba, even though Tuc Trung was quieter and more isolated.

For the first time since I had left the army, I felt the pleasure of doing what I had set out to do. I became enthusiastic about this very complete and diverse job, which included not only varied aspects of rubber production, but also clearing forests; building villages, churches, temples, pagodas, and roads; planting rubber tree seeds; and managing everything on 10,000–15,000 acres of land.

One of my most amazing experiences, which I will always remember with excitement, was my first encounter with the elephants at Tuc Trung. The plantation was surrounded by jungle and forest that stretched to the north, an area named the Hauts Plateaux. Although Asian elephants were kindly animals, they were attracted to the sweet taste of young rubber plants, which they consumed in massive quantities. Therefore, our extensions, about 1,000 acres of young plants, were occasionally invaded by elephant families that liked running, playing, frolicking, crushing, and eating hundreds of young plants. Each incursion was a disaster, and I had to try to stop these invasions. Of course, we were not savage and we did not want to kill the elephants, so I hoped to find some other solution. There were several possible options, but none of them was 100 percent efficient. For example, I tried using wardens with torches who beat on tin cans, moved about, talked, and generally bothered the elephants in order to prevent an invasion. However, these animals were very clever. They waited for the wardens to move further away, or they entered the extensions on the opposite side, playing hide-and-seek as it were. I remember having watched a funny scene one evening with my field glasses. A column of at least four elephants, including one baby, progressed in single file, each one's trunk holding on to the tail of the animal in front. As soon as they heard the tin cans, the elephants let go of the tails they were holding, made a smart "about face," and started off again in the other direction, reestablishing trunk-to-tail contact as they moved ahead.

Another solution to the elephant problem—more efficient, but very expensive—consisted of digging a huge trench around the several square miles we had to protect. This proved unrealistic because of the monsoon. It rained about eight and a half feet per year, and the rains ran level in all the enclosures one or two months of the year.

Other more or less efficient solutions had been tested by the elders but, in my opinion, we simply had to adapt our efforts to the elephants' actions. We often had to replant sacked areas, losing anywhere from a few months to one or even two years in raising the plants, according to the age of the planted area and the frequency of the elephants' invasions.

In general, there was a lot of game in Southeast Asia, and the plantations, surrounded by forests, were particularly attractive to wild animals. The country swarmed with wild oxen, stags, rock deer, boars, peacocks, rabbits, wild cocks, and more. I never drove around the plantation (about twenty-five to thirty miles) without seeing several animals. Those French who had lived in Vietnam before had almost massacred these species. Although I was a good marksman, I did not like wanton slaughter and, although I generally carried a rifle when I drove the jeep to make fire inspections, I contented myself with bagging only one animal.

One day my jeep broke down when I was at the extreme north of the plantation, on the edge of the jungle, and about six miles from the central village. I had not brought my rifle along that day, and I was alone because my usual driver was sick. I tried to find the problem under the hood, but after a few minutes I gave up and decided to walk home. I knew it would take me at least two hours to get back and that darkness would fall in an hour, but I had no other choice.

I had covered a few dozen yards when I heard a continuous noise from the forest; branches were breaking about twenty yards from me. At first I thought it was a tiger. As I was without my rifle, I got ready to climb a rubber tree. Then I saw about a dozen boars charging me from the jungle. I was so surprised that I lost a few seconds before starting to run to my jeep. I climbed on the hood (the jeep did not have a top because of the dry season). There I could breathe a little more easily, thinking that the boars were better than tigers. I counted thirteen, including a big male with two impressive tusks. He rammed the jeep with violent blows, and the others, except two or three piglets, joined in. The jeep was shaking and so was I. I must have looked like a bad tap dancer trying to regain his balance. Furthermore, the boars seemed never to have seen a car or a human being at such close range before, and they seemed to like the new experience.

Although I tried to make them go away with cries, kicks, and honks, their little game lasted until the arrival of darkness. Perched on the hood, I pricked up my ears and, when I was sure that they were gone, I got down from my perch and started toward the village. It was very dark under the rubber trees. I fell several times but sustained no injuries. I was able to think a lot about my daily life since my arrival in Vietnam. The ways of God are unfathomable. I had thought I would have some problems with communist infiltrations from the countryside; I had not thought I would have to face ludicrous situations involving wild animals.

However, the problem with communist infiltrators would come soon enough.

After about a two-hour walk, I arrived at my office, where I gave instructions to retrieve my jeep. After a shower, I had dinner: a piece of boar that my supervisor had given me a few days before. I savored each bite as I reflected on my encounter.

About five years (1956–1960) passed very quickly. During this time, I was posted to several plantations all over South Vietnam. In 1960, I got my second holiday in France. The company gave us a four-month holiday every thirty months and, of course, I appreciated the first two or three months, but during the fourth month I suddenly felt restless because I heard through the French media that the political and field situations in Vietnam were deteriorating. The Viet Cong (VC) were present in more places and in greater numbers, and more U.S. advisers had been posted to Saigon.

Upon my return, I was appointed to the Long Thanh plantation, located between the China Sea and Saigon. I felt at once that the situation, the work, and the daily life would be more difficult than they had been before. The fact was that a plantation surrounded by jungle was isolated, so the VC, who only moved on foot, came and went easily, while the South Vietnamese troops moved most of the time in noisy trucks and tanks. The plantations were jammed between two opposing forces: the Viet Cong, who were everywhere requiring taxes and food and not hesitating to mutilate or kill people, and the official governmental power with U.S. advisers and troops. The presence of the South Vietnamese troops was not strong in the field. In other words, the official government held the towns and, during the day, the main roads; the VC held the rest of the country, especially the roads at night.

My situation became a set of complex paradoxes. Although I detested the Viet Cong, I found myself in almost daily dialogue with them in order to produce as much rubber as possible. Although I personally sympathized with the Nationalist cause, I found myself cut off from their support and feigning an understanding of the communists in order to survive and function. My personnel followed my lead but with increasing difficulty as brothers and sisters became or married into either the Viet Cong or the South Vietnamese Army. Many of these people made their decisions in order to either escape prison and death from one side or prison and torture from the other. Loyalty was more a matter of survival than political ideology, and survival became increasingly tenuous.

As opposed to the good memories of my early days on the rubber plantations, the ones that followed were a lot more stressful, even awful, and are related to a long period of war in which I was involved in spite of myself. The quiet days in Vietnam were over.

Two

I Am Abducted by the Viet Cong

In 1961, I had a new but very unpleasant experience. Losing one's face (one's respect) was considered the greatest of insults among Asian people. With face went power, trust, and dignity. This cultural phenomenon placed the Viet Cong and me in a precarious situation. At Long Thanh, I was the official leader of several thousand people; the VC had to operate in obscurity. I came to understand, through conversations with some of my foremen, that the VC would not tolerate this situation. They decided to belittle me publicly in front of my personnel—who respected, obeyed, and trusted me—by abducting me.

The Viet Cong's infiltration of our area gradually intensified. Their penetration was made easier by the absence of South Vietnamese troops as well as by the Viet Cong's ability to move everywhere on foot, silently, mixed in with peasant workers. They also had a perfect and full knowledge of the people, the area, and local politics.

One afternoon, I was checking a fire lane where about two hundred people were working in a sector located eight miles from the center of the plantation. Suddenly I was surrounded by several threatening Viet Cong, their guns aimed at me. I realized with shocking clarity that while the Viet Cong had nothing to lose in this matter, I did. They wanted me to lose face. They were a dozen or so armed men and I was alone with empty hands. I braced myself for the demands that I knew they would make.

I did not try to run away before the Viet Cong got close to me because I really had not seen them, indeed, how could I have located some people in black clothes among a lot of others in black clothes scattered on several square miles of land planted in mature rubber trees that were spaced only five meters apart? Even if I had seen them soon enough, I would not have run away because they had the opportunity to appear any day, anywhere. It would have been useless to shun them. In addition, my general inclination to accept challenges pushed me to accept this one at once. Sooner or later I would have had to go through it.

Their guns pressed against my stomach and back, I got only a glance at the petrified look on my workers' faces as the menacing silence thickened. When the Viet Cong were confident that everyone was watching them and me, their leader slapped me strongly across each cheek and commanded me to kneel down at once and put my hands on my head. I had heard accounts of other abductions, but when it actually happened to me, it was very different. Nevertheless, I had promised myself that if such a thing should occur one day I would keep calm, and over all, keep my dignity. In a few fleeting seconds, I recalled my mother's death when I was only nine years old; my first glare into the eyes of a Vietnamese communist as we aimed our weapons at each other during my 1952 military tour of duty; and, ominously, the death of my good friend Albert by a bullet fired from ambush during the French-Communist War. For the first time in my life, I was being forced to do something that all of my character cried out against. Death seemed imminent. Had I cried out, rage would have given me courage, but my mouth was dry, my hands were sweaty, and my heart beat madly. In a flash of defiance, I decided to pretend not to understand the demands of my captors, who gazed up at me. A quirky sort of bravery robbed me of my rationale, and I forgot the guns sticking into my abdomen; my brain ceased to function; reflex took control. Deep inside me frustration boiled with the need to sublimate my rage.

I continued to stand with my hands on my head. The Viet Cong leader stretched up, bawled into my ears, and slapped my face repeatedly. Behind me, I could hear rifles being cocked as another VC prepared a rope to tie my hands behind my back. Later I would reflect, with some pride, that because I had not knelt down, the Viet Cong could not reach my hands on my head, and, as I continued to stand, I received two more slaps before I had time to lower my hands. At last I was tied up and, before I was gagged and blindfolded, I had the great satisfaction of seeing that my petrified workers seemed concerned about my sad situation. Every positive detail would prove vital to my morale.

Four heavy thumps to my back sent me forward as a VC on each side grabbed my arms. The VC in charge of political affairs began a long speech, in turn theatrical and threatening, sugary and violent, as silence fell over all present. His was a simple rhetoric—well known and powerful—based on clichés and threatening images. Unfortunately, it sometimes worked, especially with people who had no former experience with political strategies or brainwashing techniques.

After a while, the speaker stopped. Surrounded and supported by guards, I was forced into the forest. Our progress was long and difficult for me because of the blindfold. We must have traveled two or three hours before we arrived at their camp, where they finally removed my blindfold. In the open forest stood a kind of small bamboo hut that sort of blurred into my vision just before they thrust me into it and locked it shut. At last I could rest and think.

The hut was empty except for a small sleeping mat and a jug of water. The bottom of the door had a hole that let a little light inside. Through this hole, I heard two sentries talking about me, so I realized that they did not know that I knew their language. They seemed to be satisfied with my abduction, but they had not appreciated what they called my "resistance." These few utterances heard by chance made me feel better. Our ideologies were opposite; my sensibilities had not yet failed to function. My confidence grew even more when the political speaker himself "honored" me with his presence and addressed me sweetly in French, "You are an adult, and as an adult, you can listen, understand, and participate." His politeness betrayed his dishonesty as he broached my abduction as an invitation, expressing his concern for my needs as if he were a host with nothing to offer.

I hated everything about this hypocrite, but because his French was not fluent, I decided to argue with him tit for tat.

"Understand and participate in what?" I queried.

"Understand what I say. Participate in what I suggest."

I pretended not to understand and asked for more details about his ideas. He responded, "I'll come back to my ideas later. First, let me make a few suggestions. One is to abolish productivity pay. Then, give each worker, with the exception of trained personnel, equal pay at a higher rate. Of course, the workers deserve more fringe benefits and fewer working hours. What do you think?"

"Well," I responded, "I work for a French company that has its own rules based on European labor relations."

He began to fume. "This is Vietnam. We don't need your inferior rules."

"I don't have the power to change the rules," I said. "If I don't follow my company policy, I'll be dismissed."

"Your company need not know about these changes. This should be a personal pact. If your company dismisses you, you can follow us."

I was dumbfounded. "Your ideas confuse me. How can I work for a company and at the same time betray it?"

He countered, "In the present situation, you are betraying the Vietnamese people."

"My workers are satisfied with their jobs and living conditions," I explained.

He accused me of being ill informed but went on. "Assuming that you are right, you know they will be even more satisfied if you give them more pay and better benefits."

"Your line of reasoning is absurd," I retorted. "Of course, no one in his right mind would refuse higher pay. But such a gift would be mortal for the company and without the company, there would be no pay for anyone."

"It is not a gift. The workers deserve it. It is abnormal for them not to have the same rights of income as the company has," he proclaimed.

I appealed to him, "What about investments, construction, and economic expansion?"

"We, the Vietnamese people, will deal with those."

I am afraid that to have contested that point would have made me the "boorish guest." I remained silent.

He then became excited, mixing French and Vietnamese. I continued my pretense of not understanding Vietnamese and he rose suddenly and left, furious.

The next morning—after a hellish night caused by the combination of a hard bed, too many mosquitoes, and far too many thoughts banging together in my head—my sentries brought a black uniform and ordered me to put it on. It was, of course, too small for me, but they seemed not to care that I looked like a scarecrow. I had to go outside, where my ridiculous garb did not even foster a grin. Quite to the contrary. The Viet Cong were sitting in a circle under a big tree and all their faces were very serious. I was ordered to sit down with them, and I had to begin my apprenticeship in self-criticism. The participants in self-criticism sessions numbered about ten or twenty, with two or three political propagandists. Participants enumerated their daily deeds and thoughts as well as their wishes concerning politics, love, friendship, food, drink, and so on. The goal was self-effacement accomplished through a sort of purging of misdeeds. Such catharsis was regarded as noble progress toward more desirable behavior. Everything that I held as sane my captors held as insanity and, of course, vice versa. Another indignation that I suffered occurred the next day when one of my former workers appeared in camp and informed his compatriots of my knowledge of Vietnamese. The political VC who had become frustrated with me the day before bore down on me with his wrath. Our conflict of the day before now became the fault of my ignorance, and I was told that my "education" would require a lot of work.

I had to participate in self-criticism sessions twice daily; moreover, I had to endure brainwashing by the political Viet Cong every evening for about two hours. Brainwashing focused on eradicating all ideas, convictions, or opinions that were not Marxist. It included endless monologues with some questions thrown in here and there that were framed in such a manner as to lead to the next section of the ever-continued monologue. "Patients" were the target of questions asked in the most intimidating tone and content. If a patient resisted, the session simply continued. The brainwashers came continuously, singly, in pairs, or more, but the patient was always, of course, alone.

Between these sessions, we moved through the jungle. This constant relocation was thought by the VC to be necessary if they were to avoid being spotted by South Vietnamese troops. My black uniform was supposed to allow me to go unnoticed; however, as we passed among peasants and through villages, I stuck out like a gawky giant. It was also meant to send a derisive message: only a Viet Cong would go about in black pajamas in the middle of the day.

The two daily meals were very slight: two handfuls of rice, one little dried fish, and three or four leaves of a kind of water lily. While bolting down these frugal meals, I could not help but think about the delicious Vietnamese dishes I was used to eating at home. Such comparisons only served to sadden and frustrate me more. I had to try to eat anyway in order to survive, though the menu was hardly appetizing. I felt abandoned and depressed. I grew to look forward to arguments with the leader as they served to relieve me of my anguish and thoughts of home. At least his convoluted rhetoric was distracting.

I had a hard time getting any boiled water to avoid amoebas. The Viet Cong did not need boiled water, and they didn't want to do anything special for me. I found the solution by getting some boiled water from villagers. They prepared rice for us, and I would take some of their boiling water before the rice was added.

The tight black clothing, the arduous route, the sweltering heat, and the constant maneuvers to avoid detection by phantom South Vietnamese troops caused quite a thirst. Fortunately, I obtained a hollowed-out coconut in which I could carry my boiled water.

In the evenings, when we reached a new camp, instead of getting some rest, I had to suffer another two-hour brainwashing that was much more exhausting than progressing through the jungle. They did not want to listen to me about responsibilities, qualifications, motivation, attendance, and so forth; in fact, according to their statements, everyone was good, courageous, nice, and moral except me. This sort of politics could go on and on, and I regularly came out of these meetings absolutely broken for hanging on to my ideas.

In general, a frugal meal followed this painful session. After that, about twenty Viet Cong soldiers, the political VC, and I would participate in one of the two self-criticism sessions held daily. We would sit in a circle around a fire (if the area was quiet). I was always flanked by two guards who did not set me free at any time. We had to list all we had done during the day in order to reproach ourselves, but in my case a special rule was applied. They considered that I did not have enough to say about my day, so they forced me—through their choice of questions—to tell them as much as possible about previous weeks or months. It was derisive, terrifying, and tiring at the same time. The first day I tried to stand up, but I quickly understood that the game was fixed. My goal every evening was to hold out in order to get enough rest to be able to walk the next day.

I had no idea how long this nightmare might last. The idea of escape crossed my mind one or two times, but I forgot about it quickly because, in case of success, it would have been impossible for me to continue my work and keep my post. My escape would have resulted in another abduction and, possibly, retribution against my workers and their families. Without me the plantation would have closed.

On the eighth day, the political VC took me to the side and began discussing everything and nothing, as if in a void. He seemed in a good mood, and he announced that I was going to be freed under one condition: I had to admit that my stay in the forest had purified me morally, intellectually, and politically. This condition would have caused me to have a guilty conscience if I had accepted it as it was, but I had learned to effectively fight words with words during my ordeal.

My immediate goal was freedom and I had to find an answer that would both satisfy him and, on the other hand, allow me to be at peace with myself. I replied that I had changed and that I now understood many new things because of living in the forest. This, of course, was true. It also allowed for two interpretations:

The communist's: He has changed, and has learned many things with us. In his common form of narrow-mindedness, the political VC seemed absolutely certain that the speeches, discourses, brainwashing, and self-criticism sessions had been effective, and that I had really gone over to embrace communist ideas.

My real meaning: I had changed because I had grown more bitter. Before my stay in the jungle, for example, I had not liked the communists, but at this moment I hated them, their indoctrination methods, and their starvation diet. However, they did not know that, at least at that time, and it was a good chance for me to get away.

The political VC seemed content with my answer. Fortunately, he did not require me to sign a pact delineating my political transformation. He ordered my two "bodyguards," who suddenly became much friendlier, to escort me to the same place where I had first met them.

After a few hours walking in the jungle (my "escorts" knew the jungle well), we came to the trees' edge where I was given back my clothing. I had to perform a sort of striptease in order to return their black uniform to them. Putting on my shorts, I noticed I had lost weight (in fact, when I weighed later, I found that I had lost about two pounds a day). My shorts fell to my feet, for I had no belt, and one of the Viet Cong voluntarily cut a creeper from a tree and gave it to me. Obviously, I preferred a creeper to nothing because I had to walk about eight miles to get to the plantation and home, so I thanked him and left. After a mutual "see you later" (that I really did not wish at all), we separated and I was alone at last! A great and deep joy filled me. I was free! I began to imagine the celebration that would occur in a few hours with my friends from other plantations. It would be time to think about the aftermath of my abduction, report to my boss, and possibly adopt a new strategy later.

That evening, after having cleaned each square inch of my body to remove the unbearable smells of grime, sweat, and coagulated rubber, I had a wild night in Saigon. I then met with both my general director and my plantation boss the next day. We talked a lot about the event. They explained to me that they had tried to get in touch with the Viet Cong about me through a plantation truck driver whose brother was a VC. His efforts might have worked; they might not have. Luckily, they were not put to the test, and no one will ever know if he would have succeeded. Anyway, I was free. That was the most important thing. As I believed that the Viet Cong would abduct other Europeans in the future, I succeeded in persuading my bosses that it would be useful to anticipate such events and take preventative measures before they occurred.

I recommended that they keep only unmarried European assistants in Vietnam and appoint married ones to Cambodia or Africa, where things were quiet. This measure could prevent possible despair and, perhaps, even reduce some of the Viet Cong's pressure. The company could also demand some help from the South Vietnamese Army, but the company president disagreed with this idea because he feared showing everyone that he took a stand against the communists when his politics were supposed to be neutral. Anyway, these men in Saigon had to make their decisions; I, in the field, had to make mine. It was not an easy time for any of us.

After two or three days of rest, I began to drive and walk along the edges of the plantation to see all the people, among whom were some unknown Viet Cong related to the workers. I wanted them to understand that nothing had changed in spite of my stay in the forest. Nonetheless, I confess that my emotions were in turmoil the first times I crossed the area where I had been abducted—my stomach ached and my heart beat a little too fast. However, I had to go back. First, to keep the trust of my personnel, and second, to regain my self-confidence in order to continue working. Saving face had become the norm for me as it was for the Vietnamese. Two days after my liberation, a Saigon newspaper, *Tieng Chuong*, spoke highly of the South Vietnamese Army's effort to surround the Viet Cong and put them to flight, thereby helping me to "escape." In actual fact, I did not see any soldier's shadow in this matter; it was, once again, a matter of the army saving face—the rule of the game.

These abductions were very stressful and painful because no one knew what would happen during such an event or after it. I was abducted three other times in the next year. The process was quite the same, but the sites of the abductions were different because the Viet Cong wanted to show that they were everywhere and were not afraid of the non-communist troops who received publicity in the newspapers that alluded to my plight. The Viet Cong also chose some fields where different personnel worked in order to spread their control over the minds of as many workers and inhabitants as possible. In fact, my personnel did not need brainwashing at all because, as we said over there, the bamboo phone worked perfectly at any time and anywhere, so everybody was very well informed about everything. Besides, the workers had little choice. If they wished to continue to support their families, they had to obey VC demands.

Strangely, even the circumstances of my abductions were similar and they occurred where chance could make the worst happen. I felt a kind of routine being drilled into me time after time. I memorized the sequence of events: the pointed guns, the tied-up hands, the walk in the forest. One time to another my stress slowly decreased. Nevertheless, the quality of my sleep suffered for several years.

The Viet Cong sized me up during these months and, perhaps because they considered my behavior insufficiently obedient according to communist logic (they were always right; others were always wrong), they had to do something else to impress and upset me in order to make me better "understand." They, therefore, forced me to witness "People's Court" several times.

Three

More Viet Cong Pressure: The People's Courts

The Viet Cong had an established set of reprisals that they used against those they considered hostile. These reprisals came about in stages that were aimed at breaking one's mind through various atrocities performed before huge gatherings of people pulled from the community. These gatherings were a sort of court.

There was no way for the Viet Cong to convene a gathering far in advance because they trusted nobody, and they feared possible betrayal or exposure to the South Vietnamese Army. Therefore, a few minutes before work ended for the day, unannounced armed Viet Cong would suddenly emerge from the forest and appear all over the plantation. They would force all the workers and their families from a given village to follow them into the jungle, but only for a short distance. In about two years, I had the sad privilege of being ordered to observe about a half dozen of these People's Courts. The VC would accost me from the heat of the rubber fields, the security of my office, or the comfort of my home with "There is a People's Court; you must come with us." Of course, I was herded to the designated place flanked always by at least two VC armed guards.

My first summons to such a court sent me reeling with memories of my abduction. The same fear of the unknown crept over me. This initial experience took place in a rubber field located about five miles from the plantation center, a site isolated and quiet, secure for the Viet Cong because they could easily escape into the jungle at any sign of alarm. However, during my experiences at these courts, there was never any outside threat. The VC had total control of the area.

When I arrived at this first court, several hundred people were already present, sitting in rows. (As I was to learn, the number of people in attendance stayed at several hundred per court.) The hubbub of the crowd stopped when I appeared. I had to sit in the first row, flanked by my guards. My heart beat strongly in my chest, for I felt uneasy about what might happen next.

The audience was surrounded by armed Viet Cong. In front of us stood the Viet Cong political leader, Cau Tan, whom I had gotten to know during my abduction. With three assistants on each side, he demanded silence by tapping on a piece of bamboo. Then, he began an upside-down speech in which everything was perfect in the world of communism and pathetic in the rest. This kind of self-satisfied monologue had always left me depressed and exhausted because I knew the poverty of such rhetoric; but I could not debate with him in public. I had to sit silently as he harangued these vulnerable people who knew nothing of the outside world.

At these moments, I could not help thinking about America's objective in Vietnam: to help the non-communist Vietnamese fight against the communists. It was unbearable for me, and sometimes I mourned to see the aberrant situation. While the South Vietnamese and the U.S. military fought communists soldiers in conventional battles, more or less keeping the roads and towns safe, waves of Viet Cong partisans quietly and persistently strove to take control of the poor and more vulnerable rural people. Finally, and unfortunately, they succeeded.

I remember that I tried discreetly to talk about this situation with the local authorities, but their answers were always the same: they had no time. They pretended to listen to me, but they did not; they were polite, but they seemed not to care. It was very frustrating for me.

Anyway, after a boring two-hour speech, the Viet Cong political leader ordered the soldiers to bring the defendants. With tasteless, theatrical, and intolerable ceremony, the guards dragged in two poor young men whose hands were tied and eyes covered. The guards forced them to lie down on the ground, between the audience and the pretense of a court. When their blindfolds were taken off, everybody could see their livid and terrorized faces. A deathly silence fell on the audience.

The Viet Cong political leader hit his bamboo again and resumed his speech about revolution, communism, capitalism, Communist Party goals, Vietnamese political puppets, and the Americans. After about another hour, he explained why the two prisoners were there and how they were going to be judged.

At People's Court sessions, the charges were almost always the same. For example, a relative was a South Vietnamese soldier or married to one; the defendants had spent a while in town and met with some non-communists; the defendants talked too much. This time, one defendant had a sister married to a South Vietnamese soldier, and the other had been too gossipy outside his village! The word "criminal" echoed again and again as the Viet Cong political leader described their nefarious antics. Fear froze the audience.

Cau Tan then required the defendants to "confess their crimes" in a self-criticism session. One of them confessed that his sister really was married to a Nationalist soldier. It was true, of course, but what could he say in addition? The communists supplied what he could not. The other said he went to the market in town every week, and as everyone did, he met people and talked with them. For him, every conventional and normal Vietnamese, dressed in conventional and usual clothing, was a conventional and normal person. He thought he could say "Hi" or "Hello" and talk with people as everyone is used to doing. What answer could any normal human being give in this predicament? Nothing. So the communists did.

Cau Tan told the first defendant that he should have explained to his sister before she got married that all non-communist

Vietnamese were bad people. Furthermore, if he loved his sister, he should have convinced her either not to marry or, the best solution of all, to take a communist husband.

I mentally ran through all the arguments about love, freedom, free choice, and so on that I would have liked to give as an answer to this stupid and inept idea. However, the poor defendant shut up because he could do nothing else. The Viet Cong leader took advantage of this silence and attacked the other defendant. He explained to this one that he could not speak to unknown people at any time, in any place, or about any thing. The blood bubbled in my veins as I listened to such moronic comments, but my two guards reminded me that I had to keep quiet. The poor second defendant also had nothing else he could do but be dumb. After that, in order to save the face of his theatrical but melodramatic show, the Viet Cong leader allowed the Vietnamese audience, but not me, to speak.

The members of the audience were not foolish and did not stand up for the defendants. They knew the procedures and outcomes of these sessions, so they always kept quiet. In fact, in their minds, two victims were better than five or ten. I thought that they were, unfortunately, right. One or two sessions later my thoughts were verified when a defendant's brother got up and tried to say something in his defense. This bravery changed nothing. On the contrary, the brother himself became a suspect.

However, the Viet Cong considered a session more respectful if the community accused the defendants, or at least failed to stand up for them, as was the case at my first court. Failure of the people to stand up for a defendant was tantamount to their condemnation, so sentencing was an automatic second step in the process.

The Viet Cong leader, as judge, had a long talk in a low voice with his cohorts while the audience sat motionless, surrounded by a circle of armed VC partisans. We awaited the sentence.

The most usual sentences were having fingers, hands, or arms cut off; other times, tongues were pulled out or heads cut off. The seriousness of the sentence was in proportion to the importance of the crime, but the Viet Cong's notion of importance was completely absurd and, of course, absolutely arbitrary. Their only target at that time was to terrorize as many people as possible by threats or acts of violence—a cowardly but effective method!

If a prisoner was sentenced to die, he was allowed to smoke his last cigarette while the audience suspended its breathing. Such was the case on this day—the man with the brother-in-law who was a South Vietnamese soldier was sentenced to death. The other poor guy was sentenced to having his tongue cut out. It was a very stressful moment. Everyone was forced to wait and then watch the horrible show.

The sight was very difficult to endure. Was I awake? Where were the human values here? Where were civilization, family, love? I was almost overcome, but I had to watch.

Having finished his cigarette, the doomed man had his shirt ripped off. Then two Viet Cong laid him down on this stomach and placed a chunk of tree trunk under his head. With a single swing of a machete, one of the VC beheaded the man. The head rolled around on the ground; blood spurted all around the body; some women cried. The VC seemed pleased with themselves. Finally, the VC required some "volunteers" to bury the body at once, on the spot, without leaving any sign. At that moment, before everybody left this scene of carnage, I perceived the acrid smell of blood infiltrating the air we were breathing.

My morale was badly shaken. From then on, after such sessions, my sleep would be disturbed for several nights, and I have some staggering visual images that still remain in my mind.

Of course, the Viet Cong were known to use some psychological methods I never personally witnessed: for example, the relocation of hundreds and thousands of townspeople, forced to clear and cultivate plots of land in the jungle in order to survive if they could. These drastic acts were mixed with a lot of brainwashing. The stress, the sickness, and the despair resulted in the death of many people. Some others eventually were able to escape—many of those became the famous "boat people" who received extensive media coverage after the war had ended.

Another method consisted of putting pressure on people whom the Viet Cong thought were anti-communists: for example, the VC would threaten their families or arrest several family members. They would then release those they had arrested, only to return and arrest them again, thereby maintaining a continual harassment that took away the willpower of these people. Of course, the Viet Cong had innumerable methods for dealing with malefactors, but the methods that I observed at these People's Courts were more than enough for me. In most instances, it was transparent that the cases had been decided beforehand, although the "court" sought to display some external sense of due process.

If a defendant was not sentenced to die, the mandated punishment was executed at once. Afterward, a Viet Cong nurse would disinfect the cut limb and try to put a dressing on it. However, as the medical equipment available was very primitive and rudimentary, the tortured man sometimes died in a few minutes. If he was lucky enough to survive, he was taken away by his family. Usually, nobody saw him or his family members again. Spurred by fear, they would disappear from the area.

When the execution or punishment phase was accomplished, the VC political leader discoursed one more time for another hour or so. In turn, he talked about the dangers of collaboration with South Vietnamese and U.S. troops. It was said to be in the best interest of all in attendance to participate in the Viet Cong struggle against the bad Vietnamese and the United States. Then his tone became sugary and warm as he tried to convince everyone to give information to the VC about the movements or activities of American and South Vietnamese troops, or to help the communist effort with money, food, and/or supplies. Once he finally finished, he thanked the audience for being there and, with some more taps on his bamboo, gave the signal for everyone to go home.

After one such court, a few people went to town secretly to inform the South Vietnamese authorities, whose response was useless. After the Viet Cong had already been gone for two or three days, a battalion of South Vietnamese troops rushed to the site of the hearing, set fire to the forest, fired a few rounds, and threw a few grenades. Following this "attack," the South Vietnamese soldiers paraded through the village, noisily satisfied with their "conquest."

The impact of this sort of action on poor and ordinary people was that they often ended up helping the Viet Cong. In fact, they had no real choice. The problem for them could be formulated in this way:

(1) If they wanted to keep their jobs and their way of life, they had to stay where they were.

(2) If they stayed, they suffered under the Viet Cong, whom they saw almost every day.

(3) If they moved closer to towns where South Vietnamese and U.S. troops were quartered, they often could find neither jobs nor housing.

(4) Nationalist and U.S. troops often had things to do other than keeping in touch with the rural population.

I realized that this implacable psychological war waged by the Viet Cong had begun during the war against the French. According to communist logic, this was just the continuation of its strategy after the 1954 Geneva Conference. They also considered this a first step toward eventual Vietnam reunification under a communist regime.

The smart and opportunist VC took precautions during these actions and contacts with people. For example, they took care never to use the word "communism." Most of the people in the countryside, about 85 percent at that time, didn't know what communism was. Instead, the Viet Cong talked about nationalism; about love of country, family, and land; and about foreign invaders. In this way, they could touch the patriotism that exists in the hearts and souls of every citizen in any country.

Generally speaking, it is easier to persuade or inculcate in someone else's mind some new idea or ideology when a person as yet has no firm opinion than it is to make a person change his opinion. With the passing of years, that is what the Viet Cong did—primarily in the countryside, but eventually all across the country. As a result, a few years later, all Vietnamese, whether communist or not, whether wanting it or not, became citizens of a communist-controlled country.

The South Vietnamese regime is responsible for this situation for several reasons:

(1) Many South Vietnamese, with their ready access to jobs, money, and supplies in the economy generated by the presence of U.S. troops, had little motivation to combat the VC's campaign of psychological terrorism. Leaving the countryside to the VC made the way open and easier for them.

(2) A large number of South Vietnamese soldiers had relative(s) and/or friend(s) living in the countryside who already had been convinced or brainwashed by the VC. The temptation to remain neutral in order to protect these unwittingly involved people from VC reprisals was sometimes irresistible.

(3) South Vietnamese and U.S. military decisions concerning the establishment of strategic hamlets led to the destruction of some villages and the forced movement of villagers to "secure" areas. Designed, in part, to counter VC propaganda, this strategy failed to take into account that walls or fences cannot prevent the circulation or spread of ideas. Inasmuch as a few VC—or even one VC—had been part of a village before it became a strategic hamlet, the move only confined the "evil" inside the walls meant to keep it out. Rather than either stopping or even slowing down the spread of VC

> propaganda, this stategy—doomed before implemented —favored its spread.

I remain deeply convinced that the only way to have fought the long-term psychological pressure brought to bear on the people by the VC would have been the mounting of an effective psychological counteraction by the South Vietnamese.

Four

The Abduction of Jean Pernet

The Viet Cong launched a progressive invasion that took bit after bit of the area around Long Thanh and Ong Que, our plantations where I was originally assigned. As the situation slowly deteriorated year by year, my encounters with the Viet Cong increased. One of the most difficult meetings was trying to negotiate the freedom of my colleague Jean Pernet, whom the Viet Cong had kidnaped from Ong Que (located some eighty miles from Long Thanh, where I worked).

His scenario was much like my own. He was arrested in front of his workers and locked up in solitary confinement in the forest with mosquitoes, heat, little food, and plenty of stress. After two weeks, Jean Pernet was still a prisoner. For some reason unknown to me, the director of Ong Que plantation had not been able to get in touch with the Viet Cong about him. Everybody was worried because these events usually did not last so long. On the sixteenth day, my director summoned me to Saigon and asked me to try to get in touch with the VC around Ong Que and negotiate his release. According to him, I had both the requisite experience and the needed fluency in Vietnamese. In fact, all the flattery went unheard because I was already intent on helping my friend and colleague.

I was ready to head out at once, but my boss stopped me because he wanted to be more specific about my assignment. This kind of mission did not appear on my work contract and I could have refused to do it. The situation had to be clarified so that, in case of a bad snag, my company could ignore my involvement with the Viet Cong in this matter. Its position would have been that Mr. Michon decided to meet the VC himself in order to negotiate his own deal. I left without a word. I was a little angry because, at that time, I did not care about my contract. My only and immediate goal was my friend's freedom, and I would have preferred to hear some encouragement rather than some bureaucratic flack.

Anyway, I took my jeep and drove to Long Thanh thinking about the best method to use in order to help Jean Pernet. My task was quite hard for two main reasons. First, the two plantations were located far from each other and, between them, there were only forests—no villages, towns, or roads—just jungle. Second, my driver and confidant explained to me that the Viet Cong organization was made up of a lot of autonomous partitions, or cells, which did not know each other, except for the political leaders.

I had a dilemma. If I went directly to Ong Que through the forest, I would no doubt be arrested by Viet Cong who did not know me. Under those circumstances, I could find myself joining my colleague as a fellow prisoner rather than as a negotiator for his release. I could not risk this approach. If I tried to get in touch with the VC who controlled my plantation, they might not care about the matter or they might not want to interfere with it.

Although neither of these approaches was perfect, I decided to try the second one because there was a little advantage: I knew some of the local VC and they knew me. This may seem insignificant, but it was important for me before I embarked on such a new and unknown adventure.

Through my confidant, I made contact with two anonymous Viet Cong one evening on the edge of the forest. I had no precedent to follow. No Frenchman had been fool enough to try to contact the VC. In fact, we rather sensibly avoided them. I explained that I had been chosen to act as an intermediary between the general director of the plantations and the Viet Cong at Ong Que. I thought that a lot of explanation might make them feel like taking charge of this deal, perhaps becoming more important and responsible themselves. They listened to me, but they seemed interested in something else. After I ended my explanation, they evasively said that they were going to think about my interest in contacting the VC at Ong Que. This answer was not very encouraging but, as I had no other choice, I thanked them and went back home.

Several days passed, but nothing happened. I began to think about the "do-not-care Viet Cong way." My general director harassed me two or three times a day by our plantation radio and he pressed me to act as quickly as possible. However, I was afraid of bothering the Viet Cong if I talked too much about a subject in which the Long Thanh VC were not in the least interested.

In spite of the seriousness of the situation, I could not help laughing to myself about this strange situation. Usually, every day, every week, every month, I tried to avoid meeting the VC as much as possible, but now I ran after them in hopes of starting a conversation.

However, I had to reconcile patience, firmness, diplomacy, and anxiety while, at the same time, continuing my daily life routine in order to avoid a possible betrayal by some dismissed worker, his relatives, or his friends, to South Vietnamese authorities.

About a week later—by that time my general director was panicked—my confidant woke me up in the middle of the night and told me that two Viet Cong were waiting for me on the edge of the forest. I jumped into my shorts, then into the jeep, and drove quickly to the place I expected them.

When the two Viet Cong were sure I was alone and unarmed, they came out of the forest and met me cautiously. I had never seen these two VC. I introduced myself at once, but the only thing they said was, "We are from Ong Que. Come with us." My heart beat faster, and a joy came over me: finally I had made contact! I had to leave my jeep, and I could not take my confidant with me because the direction had "to be kept secret." The VC told me that the Frenchman was in solitary confinement and nobody could know where he was.

Their terse response was resounding, but I hoped that I had heard them wrong. Unfortunately, I had heard them right. I had to walk for miles and miles with a blindfold over my eyes. It was nightmarish. The jungle was full of bushes, branches, and tree trunks, and I was being dragged along, at night, by two unknown VC. The experience was far more tiring than any marathon.

I cursed the stupidity of the Viet Cong's orders one more time and, above all, the idiocy of their enforcement. The Viet Cong walked in the forest as if they were at home. As far as I was concerned, every tree looked the same, and I had no chance of finding my way by myself even if I had had no blindfold. I tried to explain that to them, but they remained deaf to my comments.

As we slowly progressed, I could not help thinking about my previous abductions because I was, for the moment, in the same situation. Needless to say, I hoped that the end would prove beneficial.

After several hours of walking, we arrived at a hut, where I fell down, completely exhausted. Someone took off my blindfold. It took me some time to see anything again. A Viet Cong gave me some bitter tea because I was so thirsty I could not speak at all. A few minutes later, four VC I didn't know came into the hut and we began the discussions. In fact, the VC talked in turn as I listened. At first, I had to suffer their usual boring speeches about communism, capitalism, good people, bad people, and so forth. I tried to stop them from talking by asking about my colleague, whom I had not seen yet. They said only that he had to be kept invisible and far from where we were. I was very disappointed, but had no choice but to listen on and on.

When they were satisfied with their own gibberish, they listed their requirements. All plantations were located in areas controlled by the Viet Cong. Hence, we had to pay real estate taxes to them. (Although this claim was not officially true, it was fast becoming reality. And, of course, we had to pay taxes to the South Vietnamese government as well.) I noticed that they talked only about "invitation," never about "abduction," and this demand was their way of not claiming ransom in exchange for Jean Pernet. The irony thundered inside me, but I kept still and listened.

They demanded 500 piastres per acre, about 3 million piastres per year, for Ong Que plantation alone—a huge sum. Indeed, the average salary of a Vietnamese worker at this time was about 1000 to 1500 piastres a month. When multiplied by a thousand Ong Que workers, the monthly total would come to 1.3 million piastres. Therefore, the Viet Cong required a huge sum equivalent to about two months total in wages for the Ong Que plantation. Who said that money has no smell? Anyway, I had no power to discuss or decide anything at all. I first had to meet my general director in order to transmit this news to him.

I will pass over my return, which took place in the same awkward manner and under the same difficult conditions as the trip coming out. Having spent the whole night stumbling through the jungle, I was tired, thirsty, hungry, and angry when I got home early in the morning. However, I had to leave for Saigon immediately.

In Saigon, my general director, who was with his assistant and my own boss, welcomed me with unusual excitement and assailed me with questions. My mood was not right for this interrogation, which frayed my already frazzled nerves. I became sullen. Gradually the conversation took on a friendlier tone.

When I finished my account, the general director was less excited and a little demoralized. Nonetheless, I had no doubt about his will to do the best he could to rescue Jean Pernet. The matter was not easy because of the context. The plantations were astride nonexistent borders between Viet Cong and U.S. troops (along with their South Vietnamese hosts). And there, inextricably, we were trying to function and produce rubber.

My bosses conferred in an adjoining room for a few minutes. Then the general director explained to me that they wanted me to try to re-encounter the Viet Cong as soon as possible. Although they urged me to try to negotiate for a smaller sum, they assured me that I could agree to whatever was necessary in order to gain

Pernet's freedom. I thanked them for their confidence, but bore the strain of my uncharted mission heavily.

During the two-hour drive back to the plantation, I tried to construct a plan for my upcoming discussions and negotiations with the Viet Cong. I wondered what I should tell them. I pondered arguments that might lower the demand. The VC presumed all capitalists to be rich and ready to cut a deal. First, though, I had to establish contact with them. In the process, I had to remain completely clandestine or face prosecution by South Vietnamese authorities as a war criminal while, at the same time, remaining a silent enemy of the communists.

By the time I got home, I had already decided to use the approach I had used before because I was in a hurry and had no time to invent a new one. With Pernet's urgent situation constantly on my mind, I went in search of my confidant in order to send him to meet with the Viet Cong in the most discreet fashion. After that, I fell on my bed and dived into some frightening nightmares about huts, blood, money, and VC.

I waited, working as usual, for four long days. I knew that these days were passing even more slowly for Jean Pernet than they were for me. The communists were in no hurry. By the time Pernet's imprisonment reached a month in duration, my confidant approached me at about 7 A.M. to tell me that two VC were waiting for me at a different place at the edge of the forest.

I immediately headed out and met the same two Ong Que Viet Cong as before. In spite of my explanations and protests, the same procedure with the blindfold was followed. However, the walk on this excursion was shorter and, when they removed my blindfold, I felt a little easier than before. Still, I was suspicious of my hosts' intentions.

The same four Viet Cong I had met with before came into the hut where I was waiting. After listening to a new, shorter speech, I began explaining my general director's proposals. Their faces remained expressionless, so I was not sure if I was moving in the right direction or not. I had to come to terms with that silence. I tried to explain that the money they required was so much that the company would risk bankruptcy meeting such a demand. After three hours of discussion and negotiation, I heard some familiar jangles outside. My stomach contracted with anxiety as I contemplated the weight of the shackles placed around the limbs of prisoners. The VC studied me as I listened to shouts and orders being issued outside. I regained some confidence, sensing that the histrionics outside were being performed to impress me in some way. I continued my appeals for some while. Suddenly, the VC stood up and left.

I waited, worried, wondering what was going on, for about two hours. Locked inside the hut, I sipped bitter tea. The Viet Cong were once again proving their flair for playing with people's nerves, a fundamental strategy in their psychological war with me as well as with South Vietnamese and U.S. troops. I half expected to be thrown into prison alongside Jean Pernet, or worse, into solitary confinement as part of a plot to increase the ransom. In alternate flashes, I anticipated Jean Pernet's release with joy. These opposing scenarios played across my consciousness, carrying me from one emotional extreme to the other and testing my patience—as the communists knew it would.

When the four Viet Cong returned, unruffled as usual, my blood began to flow faster through my veins. I scrutinized them carefully. They endeavored to review me on their policies concerning communism, capitalism, good people, bad people, and so on. They explained their disappointment with my general director's attitude and their appreciation for Uncle Ho's (Ho Chi Minh's) leniency. That meant that the VC had accepted our offer of 2 million piastres in banknotes to be paid as soon as possible. Although they did not want to hear the words "abduction" or "ransom," I had to remind them that my mission included bringing Jean Pernet back with me as well as meeting their demands. They answered that it was possible inasmuch as their "invitation" was coming to an end. I chose not to comment.

While they were talking about Uncle Ho's leniency, I understood that, although they would not state their feelings clearly, they expected me to refuse to pay. Indeed, their ideology forbade them from thinking about an exchange of a mere man, who was worth nothing, for money, which was the lifeblood of the war effort. Their belief was that a man should sacrifice himself until death rather than make his nation or friends pay for him. This rationale revealed their—to my mind, inept—concept of human values. Unfortunately, the French and, later, the Americans never fully understood the power of such an ideology during this civil war. In this atmosphere, I, or anyone, was just another pawn in a conflict that was to result in communism's victory. This slow, psychological defeat of the enemy was followed by a swifter, easier military defeat.

After hearing these four VC out, I agreed with them that we would meet again a few days later. However, in order to avoid possible betrayal, they did not want to specify a meeting time even one day in advance. They did not say so, but I imagined that they did not have total faith in either the company or me.

My return to the plantation was made without any problem except for the painful, blindfolded trip. As soon as I got home, I radioed my general director to say that I was on my way, and left for Saigon.

He welcomed my announcement of the new sum with obvious relief, and he invited me to have dinner with him that evening. As the roads were not safe at night, I planned to go back to my plantation the next day. The general director needed this time to collect 2 million piastres without causing bank or nationalist authorities, who harbored some suspicions about the French plantations in the VC zones, to become troubled. Accomplishing that was my boss' matter, not mine. Conveying the money and carrying my mission through to a successful conclusion would soon start and prove much more difficult and dangerous.

On my way, I was a little worried because being alone in a jeep with several bags crammed with 100 piastre banknotes did not seem very prudent. On some roads it would not be safe at all. Thinking what could happen to me, I pushed down the accelerator, trying hard not to think any more until I got home.

There, a new problem presented itself. Since the Viet Cong did not want to fix an appointment in advance, and since the plantation safe was too small to hold it, I had to live with all this money close to me until I could get in touch with the VC and they could decide when we could have a meeting.

The tension mounted. As the middleman between the Viet Cong and my general director, I was responsible to two parties for this money. If anything should happen to that money, the odds were that I would spend the rest of my career in Vietnam owning only a pair of underpants.

That night, with all the bags of money under my bed and my hunting rifle in my arms, I slept very badly. Fortunately, the next morning, the Viet Cong, lured by the money, contacted me through my confidant. I had an appointment at a specific place near the edge of the forest and I delivered my bags by jeep. There, six well-armed VC took charge of the bags and of me, blindfolded of course. We began our walk through the jungle to a place unknown to me. My fear, intensified by anticipation, began to mount—maybe because I worried too much, but this I could not help. In point of fact, my situation was not very comforting. The bags containing the 2 million piastres were being carried by six armed communists whom I could not see at all. It was a very disquieting situation. Anything could happen.

A little later, I realized that I was being too pessimistic. After a walk shorter than before, I took off my blindfold to find myself in a little clearing where a company of Viet Cong, about 120 armed soldiers, were camped with their red star flag flying in the wind with impunity, alarmingly close to South Vietnamese and U.S. troops. I was somewhat impressed, I had never seen a Viet Cong flag flying like that. The only other one I had ever seen had been captured in North Vietnam during the French-Communist War. We had prized it as a trophy.

Due to the Viet Cong's penchant for theatrics, they wanted to give a specific meaning to the event in which I was involuntarily going to participate. They had invited, or forced, the surrounding village chiefs, VC leaders from other areas, and several people about whom I knew nothing (but who were, I guessed, sympathizers charged with spreading propaganda) to attend.

At this stage, it occurred to me that a rubber plantation paying tribute to the Viet Cong set a dangerous precedent. Knowledge of it was sure to spread among the locals, delivering an irreversible impact. These kinds of scenarios repelled me, and I asked the VC leader whom I knew, the one who had walked with me in the jungle, if I could stay apart from their ceremony. I stated that my purpose was to give them the money and go back home with my friend. He seemed offended by my request, and replied that it would be impossible. I had to stay and watch.

I was very angry because I felt I had been tricked, but once again I had no choice. Although I had reached my limit with their pettiness, I had to comply in order to assure the success of this delicate venture.

While the Viet Cong leaders were talking in turn (again and again repeating, as always, the same concepts), I began to think, maybe too much, about my situation in the camp in relation to my walk through the forest with those bags full of money. Since all Vietnamese, North and South, resembled each other on the surface, complicating one's ability to distinguish friend from foe, I could be denounced by an anonymous non-communist infiltrator who witnessed any part of these transactions. Ironically, I faced the possibility of being imprisoned by the very people with whom I shared an ideological allegiance.

Eventually, I dozed until a trumpet awakened me. The company of Viet Cong had begun to march around the clearing. I wondered at first if I was awake or dreaming. Once assured of my own lucidity, I worried about how all this was going to end. I hoped that the VC would not require me to solemnly present them with the bags of money in front of their audience. Thinking about that possibility, my heart dropped straight to my knees. Fortunately, and much to my relief, they spared me this humiliation.

They contented themselves with piling up all the banknotes on a small table they had carried there and requiring the audience to march past in order to see it. Repeating their ruse about the real estate tax, they said nothing about Jean Pernet in front of their audience, keeping him hidden from view during their ceremony.

I had to watch the company of VC march around the clearing again. Then came the preemptive knocks on the bamboo and the audience was dispersed. When everyone had gone away, I approached the Viet Cong leader and asked him about Jean Pernet. "Where is he? Why have I not yet seen him?" Talking to me was of no interest to him at that time. He merely pointed to a small hut located on the side of the clearing. I ran there immediately and asked the two sentries to open the door. I was very affected when an unknown white man flowed into my arms calling my name. We embraced each other for a few seconds, then I looked at him carefully. He was Jean Pernet. I had not recognized him at first because he had a long beard, had lost weight (at least twenty pounds), and was very pale. During his one-month stay in the forest, he had suffered the same treatment as I, but for four times longer. He was very tired, all the more so since the VC had forced him, for security reasons, to change camps three or four times a week.

Our walk back to the edge of the forest took a little longer even though neither of us, that time, were blindfolded. I had to help Jean along. We discussed many things and our conversation overflowed into joy. The two VC escorting us showed their impatience. My driver/ confidant, who had been waiting for us for several hours, was there when we arrived at the edge of the plantation. The two VC walked back into the jungle while we got into the jeep to drive home. Once there, greatly relieved and toasting the moment with scotch, we savored the outcome: I had succeeded in my mission and Jean Pernet had regained his freedom.

The general director, whom I informed by radio, invited us to Saigon that same evening to discuss the matter and celebrate the event. Thanks to a mild alcoholic haze, our boss heard only what he wanted to hear.

Although this situation ended happily, the overall position of the company was not enviable. We suspected that this kind of problem was just beginning. We had to solve several crucial questions:

(1) How do we continue to work and produce rubber in spite of all these stressful ups and downs?

(2) How do we keep the company's patrimony safe through the war?

(3) How do we keep in touch with and have good relations with the South Vietnamese authorities, with whom we share the same ideals?

(4) How do we pay and deal with the Viet Cong, whom we detest, in order to continue to be able to recover our employees and do our work?

The company had to resolve these dilemmas in order to continue to produce rubber. However, to do so, it needed to take into account so many uncontrollable components that it seemed a slim hope it could be accomplished. Even so, an unshakeable commitment to go through all this swelled within me. At the time, all of us hoped the situation would be solved—some day, some way—to the benefit of all concerned.

Five

Rubber, Robbery, and the Viet Cong

In Long Thanh, as the Viet Cong's infiltration increased, more and more sympathizers followed them. The leadership of the local South Vietnamese authorities was more and more threatened. I felt they were caught in a sort of administrative yoke that prevented them from acting spontaneously because they were limited to following preplanned strategies. It was their Achilles' heel. Because of this, South Vietnamese troops were not able to intervene as often as either they or we needed.

In some areas of the country, caught between two political regimes struggling for control, incidents of robbery, pillage, and other crimes increased. We sometimes were directly involved in one of these plagues. I learned, much to my surprise, that when that happened it was quite impossible to avoid outside intervention.

One day I noticed that the quantity of rubber yielded had decreased in varying parts of the plantation. These decreases were following the sequence of harvesting. I checked with my staff to see that everything was working well both with the workers' attendance in the concerned areas and with the process of receiving the rubber at the plant. The process seemed secure in each successive phase and, in my opinion, the decrease could only have been caused by thieves who knew the bleeding process of the rubber trees well.

The rubber was irresistible to thieves. There were few South Vietnamese police, and those few were afraid of the Viet Cong. As far as the army was concerned, the soldiers had other things to do. Therefore, forced to take measures to protect our interests, we recruited our own militia.

The problem was that no Vietnamese civilians wanted to do this job because they were frightened of the Viet Cong. Having talked with a friend who had already worked with them, I decided to recruit militia members from among the Nung. The Nung, who were neither Vietnamese nor Chinese, came from along the northern reaches of North Vietnam and were old friends of the French. Tall and strong, they cared little for the Vietnamese people. They were exactly what I needed. I sent an emissary to the village and recruited eight of them. After a while, they became an efficient, harmonious, and trustworthy group. Of course, we were not allowed to have guns. Should the Viet Cong have come upon any of us armed, we would have been fired on without any questions being asked. Under these circumstances, my Nung militia got the idea of using old bicycle chains as weapons.

This team, lead by a Nung who had formerly been a corporal in the French army, managed to arrest several thieves over the next few months. We turned these over to the police, who preferred to get them that way rather than having to go out onto the plantation or into the forest in order to catch them.

One day, one of my team leaders indicated to me that he had seen some silhouettes, late in the afternoon, in an area not bled that morning. Since the Viet Cong came only in the areas being bled, where they could meet and indoctrinate workers and their families, I was pretty certain that these suspicious shadows were thieves. I decided that we had to go and check out the area.

I immediately picked up five of my militiamen and drove close to the zone about which we were concerned. As soon as we arrived, we jumped out of the jeep and went under the rubber trees. There, each of us walked along a row of trees. We were six, progressing on a width of about thirty-five yards, checking the cups and notches on each tree.

After covering several thousand yards without finding an anomaly, I decided to keep going in the alternate direction of the next day's harvest, each man following his own row. After another few hundred meters, we began to see, here and there, trees with empty cups and pieces of fresh coagulum, or coagulated latex, on the ground. The thieves knew their job very well. They chose to rob us of coagulum from the cups because it was easily transported. Another method would have been to rob alternate harvest areas around the plantation, as the three major areas rotated daily. This second method would have required transporting liquid latex through the jungle. Given the weight, such a task would have been impractical.

I ordered the men to be careful and to progress as silently as possible, holding their chains at the ready. The use of the chains was easily justified because, a month earlier, one of the militiamen had been seriously injured by three thieves wielding razor-sharp bleeding gouges.

The trail of the coagulum led us up to the forest, where we hesitated, thinking of the possible presence of trigger-happy Viet Cong. This was where prudence would have called an end to the search. My militiamen looked at me for instructions.

The immense expanse that had fallen prey to the thieves at that time angered me. Not wanting to be the victim any longer, I led my men into the jungle in search of our quarry. The further we progressed, the more I doubted my sanity. However, I could not resist the urge to keep going. After a few moments, we came across a huge heap of coagulum stored under bushes in what appeared to be a staging area used by the thieves until they could haul their booty away in ox-drawn wagons. I was elated because we had found a lot of coagulum, about five or six metric tons. Suddenly, I realized that we were only six guys in the middle of the forest. Any number of armed thieves or Viet Cong could be in the area. After all, we had intruded into their element.

I decided to return to the plantation where I could get enough workers to move the coagulum. As we were crossing the trail between the forest and the plantation, we were accosted by a score of Viet Cong who were waiting for us, guns at the ready. The problem was that none of us knew each other, so everyone just stopped short. My militiamen had their hearts in their boots and, although I was anxious, I began a rather extended explanation about why we were there. Eventually, with nothing more to say, I stopped. The VC had kept quiet during my monologue and, after I stopped, their silence continued. We did not know how to take their quiet stares. After lingering a while in their mute leers, I decided to try to leave. They insisted that we stay.

While they consulted one another in low voices, my imagination whirled with images of their plans for us. Suddenly, the guns came down. The VC leader came forward and began a speech that contained all the stupid ideology I had heard before. When he finished, he said that he had heard of me from his comrades around Ong Que, and that he and his soldiers wanted to help me find the thieves. His declaration floored me.

I replied that I appreciated the offer but that I was accustomed to acting alone with my militia, and that I could manage without their help. What I did not want to explain was my fear of compromising the plantation with the South Vietnamese authorities by working in the field with their enemy.

The Viet Cong leader did not accept my position. He explained that inasmuch as the plantation belonged to Frenchmen, it would soon become the property of the communists. Hence, he was compelled to adequately protect their future interest in the property.

I was dumbfounded by this response. As I tried to gather my wits, they prepared to join in my investigation. In spite of their thundering declaration to help, it was apparent that they did not trust me. They wanted to see for themselves all that we had found. So twenty Viet Cong, six Nung, and a Frenchman lined up, entered the rows of rubber trees and, progressing along a 150-yard sweep for three miles, checked the cup on each tree.

Stunned, I walked alongside the VC feeling as if I were in a bad dream. My militiamen checked one another with knowing eyes that were hard and worried. I was incredulous at the prospect that the Viet Cong objective and mine could be the same. The political theorists in Paris never could have perceived of such an incongruous gathering of forces.

When we returned to the huge pile of coagulum, the Viet Cong seemed more relaxed. They now had proof that we had told them the truth; there were rubber thieves in the area.

As night drew on, my militiamen and I became a little nervous. The thought of being with the Viet Cong at night did not delight us at all. I thanked the VC leader for his help, although I felt it was no help at all. Before we started to make our way out of the jungle, he explained to me that he and his men were going to investigate the surrounding villages and that they would keep me informed about their results. As I heard these words, I thought that he was insincerely trying to impress me with communism's efficiency, conscientiousness, and loyalty. I expected him to forget the whole thing and do nothing.

Finally rid of our unwanted escorts, we returned home. A few days passed without any news about this matter. Then, one evening, at about 6:00 P.M., an employee came into my office in a panic and told me that the VC were waiting at the forest border for my immediate presence in order to talk to me about the rubber robbery. I picked up the same five militiamen and drove to the appointed location.

Once there, we met about a dozen of the same Viet Cong who had "helped" us before. Their leader announced to me, with a smile that managed to be both sugary and sardonic, that he had good news. I became worried about his news. Since I had been in Vietnam, I had never heard any good news from the VC. I felt uneasy, and an indefinable and harrowing sentiment slowly came over me. I knew I did not have the same point of view as they did about what was right or wrong.

With a gesture, the leader indicated that we were to follow them into the forest. After walking a few minutes, we reached a clearing and met up with the rest of this squad of VC. There were about eight of them standing beside another huge pile of coagulum and, when we went up to them, we discovered behind the rubber two Vietnamese men on their knees, hands tied behind their backs. Obviously, by the way they were rolling their eyes at the Viet Cong, they were horrified. For a moment I felt soothed, thinking that I had misjudged the VC; they had really done an efficient job. Unfortunately, the matter did not stop at this point for, when I told the Viet Cong leader that I would take the two thieves with me (naturally not saying anything about wanting to hand them over to the South Vietnamese police), he refused point-blank to hand them over. He explained that, as we did not know how to exercise justice, he would handle everything.

At this moment, I felt panicky, remembering in a flash all the murders and mutilations I had seen in the past. Before I had time to do anything, the two thieves had five fingers each cut level with the palm of the hand! Damn! Nauseated and enraged, and forgetting the situation I was in, I burst out against the VC leader, bawling him out as I would have done in the security of my own office. I no longer hated the Viet Cong; I abhorred them. Realizing that I had been made a party to this tragedy, I became raving mad. I felt responsible for the trap that now ensnared me as well as the two accused thieves. I was not even sure that these men were guilty. For all I knew, the VC could have caught two nationalists, or any non-communists, and cut off their fingers only in order to show me, the Nung militiamen, and the local inhabitants how efficiently they kept their word.

When one of the VC had finished putting a tourniquet around the wrists and a bandage on the stumps of both prisoners, I firmly requested to be allowed to talk to them. I needed to know the truth in order to relieve my conscience a little. It was my rubber that had been stolen, and I had the right to examine the evidence. Again, my request was flatly refused and I was told to go home. This sudden injunction, implying a need to get rid of me, plunged me again into doubt about these two men actually being the rubber thieves.

A few days later, one of my Nung militiamen was found with an arm cut off, lying in a pool of blood, dead. Nobody ever admitted being responsible for the murder and, unfortunately for the plantation, all his friends resigned and left the next day.

With the help of my driver/confidant and a few volunteer workers, I became my own antitheft device. We did not have any more problems with rubber robberies for several months. I guess even if the two men punished had not been guilty, the news of their ordeal frightened the actual thieves off for a while.

The bamboo phone went into action after the mutilations and subsequent murder. When the South Vietnamese authorities heard about the incident, they sent me a summons. I had to report to them and confirm that I had met with the Viet Cong during an antitheft patrol. I was unsure how they would react and, the best defense being a good offense, I seized the advantage by saying that I was obliged to run after thieves because the police either would not or could not. This argument reaped only silence, but I was allowed to go home without any further trouble.

Nevertheless, I remained worried about the Viet Cong. Sooner or later, they would hear about my visit to the South Vietnamese and would have a reason to hold me in contempt for betrayal. In order to avoid this complication, I left my summons lying on my desk. I even stuck one copy on the notice board at the office, hoping that the bamboo phone would spread this information to the appropriate people. My plan undoubtedly worked, for the VC never said anything about it. But, my God, how my emotions boiled!

A few months later, while in France on vacation, I heard that the dead body of my driver/confidant had been found in the jungle. In addition, my substitute, Andre Roucoule, had had an arm cut off while fighting with thieves. Those who were robbing us were taking out their revenge for Viet Cong actions on us.

Six

The Slaughter of the South Vietnamese Military Post at Long Thanh

From 1960 to 1964, courts, abductions, robberies, and general mayhem recurred often as the situation continued to worsen. My experience so far left me pessimistic about how the war would end. South Vietnamese troops continued their war from afar, failing pitifully in their efforts to thwart the Viet Cong. I felt that any direct intervention on the part of the Nationalists would come too late to be effective. If it did become necessary for me to leave the country, I would do so with a bitter taste in my mouth.

Morale among the South Vietnamese was low. This lack of motivation was not shared by all training personnel, however, and one day, in 1963, the colonel commandant of South Vietnamese forces in our province decided to create a military post on the plantation at Long Thanh. He wanted to try to reduce the Viet Cong's activities and recover control of the situation on the ground. My general director, local non-communists, and—especially—I, all agreed that this was a good idea. I hoped that this post would allow us to work quietly and normally after the ordeals we had endured. We all urgently needed a boost to our spirits.

Nevertheless, my optimism did not overshadow my pragmatism. I knew that establishing a South Vietnamese post where the Viet Cong operated daily would not be an easy task. My experiences in North Vietnam during the French-Communist War had taught me the perils of trying to quell the communists.

I decided to meet discreetly with the colonel in charge in order to give him information that I considered important: the ways in which the VC interacted with plantation villagers, their methods of indoctrinating the population, and so forth. I tried to fully explain my hopes, experiences, and worries. I also advised him as to certain precautions that I felt he should take when he went into action. He listened silently to me and, after I stopped talking, offered me tea and thanked me for my advice.

I went back home with mixed emotions. I felt my visit had been useless. As usual, the military had little use for civilian advice. I resigned myself to accept the premise that the South Vietnamese colonel whom I had just met knew his job as well as I knew mine. In any case, convening the meeting had allowed me to find peace with my conscience.

For a few days, nothing special occurred; but when the military requested aid in the form of the use of an unoccupied building, I became nervous—as did my general director. I thought that there were at least two alternatives to placing a military post on the plantation. The army could provide enough soldiers to protect the central village and factory. In this plan, the village, offices, houses, mechanic's shop, car and truck depot, and all plant operations would fall inside an army perimeter. A second option would be to establish a post in a clear site, outside the center, that could be more easily defended, especially if the army could not provide enough troops to protect this area well.

However, the South Vietnamese army dictated a third alternative, more frugal in time and money, and with the fewest potential problems (at least, at first glance). In this plan, about forty soldiers would camp in a small building loaned by the plantation, around which they would erect a rampart made of sandbags.

Of course, we had no buildings that were located in a clear site. All we had was an old depot that was located in the central village—close to factory, offices, school, and infirmary. I tried to explain to the young lieutenant in charge that a Viet Cong attack when all the inhabitants and employees were on the site would have devastating results. He seemed uninterested and, as my general director finally agreed with the colonel, we loaned the depot to the troops.

The soldiers requested the daily presence of several dozen workers to dig a trench around the depot, fill hundreds of bags with soil, and then pile the bags inside and all along the trench to make a crude kind of bunker. The soldiers went into Long Thanh each night to sleep until their post was completed. We were faced, therefore, by South Vietnamese soldiers by day and Viet Cong by night, and the workers labored fearfully, dreading VC reprisals. Indeed, we did not know on which foot to dance; once again, I had to improvise and adapt my behavior to the evolving situation.

After about a week, the rampart was finished, and forty well equipped and armed soldiers set themselves up in the old depot, which was dubbed "the post." The villagers became even more cautious, and I wondered how long such a paradoxical situation could last. I knew the VC were continuing to come into the villages to get rice and a little money and, overall, to keep in touch with the people and show them that the military post did not change anything.

This incredible situation lasted about two months, during which soldiers, inhabitants, and VC behaved by day and by night as if their situation was normal. No one group paid any attention to another and, in this way, the days passed without any trouble.

When they first arrived, the soldiers patrolled regularly, checked people's identifications, and did the kinds of things that police and soldiers do during war. After a few weeks, however, the patrols and identity checks became less frequent. I watched the soldier's attitude and behavior change appreciably. I felt they were adopting a complacent outlook that was not at all compatible, in my opinion, with such an explosive situation. Angered by this behavior, I decided to invite the young lieutenant to have a drink in my home so I could talk to him about the matter. Smiling, he listened to me and, after two scotches, laughed at me and left.

I felt more angry than I had been before our conversation. I was sure the lieutenant had a chickpea in his head instead of a brain. He was a bloody young fool!

Anyway, I could do no more than watch and work. Obviously, the soldiers, who had not seen the Viet Cong for several weeks, had become overconfident. They spent most of their time playing guitars and chasing girls in the villages. Unarmed and stripped to their waists, they would leave their post singing songs and laughing at each other.

The Viet Cong, no doubt, realized this state of affairs. For one thing, they could get information from the villagers. Moreover, as no one looks more like a Vietnamese than another Vietnamese, I was quite sure that a few VC had infiltrated the villages, where they rubbed shoulders daily with the soldiers. Furthermore, the "truce" being observed by the VC, plus the absence of any People's Courts during this period, seemed suspect to me. I began to think that I had been right about the suggestions I had tried to make to the lieutenant. I felt I had to try to talk to him one more time, although I was embarrassed by this hard but necessary step. This time he checked me bluntly by saying that he knew his job and that I was there to produce rubber, not to wage war. This was true, but I was quite upset.

A few days later, at about 6:00 A.M., I was in my office with my staff, checking the daily attendance, when a sudden fusillade of rapid automatic weapon fire and grenades erupted near us. Everybody threw himself down at once and I crawled to a window to see the direction of the shooting. The Viet Cong had made a surprise attack on the post!

We held our breath and, in less than thirty minutes, the shooting stopped. When I looked at the post again, I saw some VC coming out. Others, surrounding the post, had their weapons pointed at windows and doors. If the Viet Cong were walking quietly out of the post, it meant that the soldiers were either captive or dead. At last, one Nationalist soldier, his hands on his head and with a gun pointed at his back, emerged. It was a disaster, and I had a lump in my throat thinking of the fate of the others. I felt hopeful again for a few seconds when some shots cracked inside the post; then more VC, looking triumphant, came out. They had finished off the wounded soldiers.

I had no further time to think about them. The Viet Cong, no fewer than a hundred men, spilled into the village, the offices, the factory, everywhere, and gathered all the people without exception: men, women, children, old people, and me—about two thousand people. Then they took us away to the edge of the forest. As we walked through the rubber trees, the VC picked up more workers. We kept moving for about an hour. Nobody dared to speak. I found something staggering and incredible in looking at this poor, sad column that snaked along in silence.

Having penetrated into the jungle a few dozen yards, the Viet Cong ordered us to sit down along a trail as close to each other as possible. I wondered what was going to happen and, although I was a little overwhelmed by the recent events, I was not too worried about us because of the great number involved in this situation. We had hardly sat down when we heard a trumpet playing the *Internationale* (the communist anthem). At the same time, the South Vietnamese soldier had his head cut off in front of the audience. Next, we heard a march in quick time just before we saw a march-past by the Viet Cong in close order.

Although I was torn by a powerless rage and deep gloominess, I automatically counted the rows of Viet Cong lined up three by three. There were 180. I had never seen so many grouped together. I understood in a flash why the VC were forcing us to watch their parade. The more who saw, the more there would be to tell what they had seen. The VC sense of theater and production was paying off for them one more time.

This macabre sense of production confronted the crowd again as they were being returned to the village. In front of the razed post, the Viet Cong had stretched out the Nationalist soldiers' nude bodies, pathetically aligned along the road with a few broken guitars thrown in. The VC who were escorting us explained that the "People's Army" (as they referred to themselves) needed the uniforms and weapons of the nude soldiers.

I could have wept, and I could assess the extent of the damage that such a spectacle had on the minds of these poor people who were not prepared to be confronted with such horror. The VC ordered the villagers to take care of the bodies. Everybody looked at each other. The villagers realized that they could not bury the bodies without notifying the soldiers' families and, as nobody knew them, there was only one solution: the South Vietnamese authorities had to be informed. As no one in the village dared to do that, all eyes were turned to me. I decided I would get in touch with the South Vietnamese as soon as possible.

The Viet Cong leader approached me and explained that they did not have a choice. They had to burn the post after their victory. After all, he said, the plantation, unlike the VC, had money to rebuild the depot so that the plantation would be kept intact for the communists, who would inherit it after the war. What he said rankled, so I turned my back and went away without saying a word. It was risky, but it worked. I was fed up with listening to the same speeches, and I did not stay to listen to their terrorist propaganda. For once, I was allowed to do what I wanted.

Afterward, I went to the South Vietnamese Headquarters in Long Thanh to inform the colonel about the plantation post. He flew into a white rage, as if I had been responsible for the disaster. I responded with silence and, as I had with the Viet Cong, turned and left his office.

About two hours later, at least a battalion of South Vietnamese troops, some 800–1,000 soldiers, turned up, paid their last tributes to the dead soldiers, combed the whole plantation, shot hails of bullets, threw some grenades, and left without seeing any VC. Depression crept over me as I pondered the turn future events might take.

"Why did the post fall so quickly into enemy hands?" I asked myself continuously. My driver/confidant gave me my answer a few days later. Several times a week, a few girls would spend the night at the post. The day before the attack, a Viet Cong girl had joined the others. She had opened the door early in the morning so the attackers could enter.

After the carnage, the South Vietnamese authorities decided to construct a trench with a dirt embankment around our central village (some sixty acres). Villagers were conscripted each day to do the actual work. Family members dug all day and, after performing their daily duties, plantation workers labored on the wall as well. Although the military came every morning to supervise the building site, they went back home every evening. The villagers' motivation was not great; everyone regarded this wall as totally useless. The fact was that a huge rampart, even one surmounted with barbed wire, was of no use if the four entrances were not guarded and controlled both day and night. I suspected that the soldiers agreed, but they were motivated by the need to save face in an area for which they were responsible.

The construction of the wall took about three months. Then the villagers were told that they, and not the soldiers, would have to guard the wall, as the soldiers would not be staying in the enclosure after it was finished. We were going to become a "strategic hamlet."

The strategic hamlets policy had several goals, including:

(1) encouraging the people to defend themselves against the VC,

(2) taking a stand against communist ideology,

(3) making it difficult for the VC to keep in touch with the general population, and

(4) avoiding U.S. bombing by the flying of a yellow- and red-striped Vietnamese flag.

I have already discussed my views on the ineffectiveness of this plan. Over time, I progressed from regarding the South Vietnamese army as pathetic, to seeing them as stupid, infantile, and, finally, unbelievably inept. Although I did not want to heap criticism on the Nationalists, this was—nonetheless—my reality; my actual experience. Perhaps the Viet Cong were able to motivate the people more because they themselves were more motivated. In any case, at the time, not knowing how this would end, I continued to hope for a positive outcome. Unfortunately, in 1975, we witnessed the opposite result.

However, life was going on, and some time after the completion of the wall of hard work and dirt, the South Vietnamese military heard about the Viet Cong who continued to routinely come into the village almost every night. The army conceived the ludicrous idea of shooting some mortar shells onto the soccer field that was supposed to be part of the nightly route taken by the VC. I did not appreciate this incongruous and completely useless initiative, especially since the soccer field was located about a hundred and fifty feet from my house!

I went the next day to visit the colonel in charge. He was not in, and I was received by his assistant, a captain. I explained that one mortar shell had fallen about fifty feet from my garage and that I was going to become a man more dead than alive if the shooting continued. Like his colonel, he remained enigmatic, and then he asked if I had some information for him. I told him about my feelings on the matter of the attacked "post." He made no comment, so I went out. Over the following days, mortar shells continued to explode, but on the far side of the soccer field. It was, at best, a gesture of compliance, but my sleep continued to take a hammering. In the morning, we would fill up the holes and, at night, the military would make new ones.

Such a feeble response on the part of the South Vietnamese caused me to doubt their ability to perceive the war as real. In my world, people were dying every day, and distress and fear was everywhere. Finally, in the midst of growing chaos, it became the onus of U.S. troops to provide security, which they did with a vengeance that turned nightmare into searing reality.

My father and two of my uncles served in the French army during both WWI and WWII. I had been brought up to admire and respect the United States for its efforts during both World Wars I and II. My generation grew up with the words of French singer Michel Sardou, "If the Ricains [Americans] had not been there, we [French] would be somewhere in Germany." Today, with the passing of time, I can look back and judge events in Vietnam better than I could when I was living through them. I can now say that the United States showed good intentions and great courage under the circumstances and conditions of the time as we now know them to have been—even though the war itself was lost.

Seven

Marriage and Reassignment to Cambodia

From 1956 until 1963, one could observe South Vietnam's economic and political degradation—slowly at the beginning, then faster in the final three to four years. The 1963 assassination of President Ngo Dinh Diem did not help matters; neither did the leadership of the four generals who succeeded one another over the following years. The communist's penetration into the country, and their unrelenting pressure on the people of the country, became inescapable.

In my opinion, the Vietnamese Nationalists did not play fair with U.S. authorities, who were, therefore, unable to accurately assess the crisis. If the South Vietnamese had convinced their allies of the potential danger inherent in the situation, the United States might have offered more effective help. However, the South Vietnamese—either through sins of omission or acts of ungratefulness—failed to inform the United States of the psychological war that had to be fought and won before any conventional war could be successfully waged. South Vietnam authorities could have won this psychological war in the early sixties; the intervention of the United States could not.

Because of this lack of candor, one was left, on the one hand, with political and military authorities in South Vietnam, especially Ngo Dinh Diem, who hated the communists and who had an unshakeable will to get aid and participation from a country as large, important, and rich as the United States. On the other hand, one had U.S. authorities, representatives of a country known to champion democracy and freedom, who thought they could not ignore South Vietnam's cry for help. The consequence was the Vietnam War.

During these years, the pressure and the tension also increased on the plantations. Consequently, the production of rubber became more and more difficult. The harvest teams' output, the factory operations, the food supply, and energy productions were constantly in jeopardy. The plantations kept going, but their future was doubtful.

After my third thirty-month stay in Vietnam had been extended three months, I went to France for a well-deserved, four-month holiday at the end of 1963. In the months preceding this vacation, I had met a young Eurasian girl in Saigon and we had fallen in love at first sight. We decided to get married at once, and none too soon as I was already thirty-three years old. I had forgotten that my company considered married men to be at risk in Vietnam, and I was surprised when my general director, having learned of my marriage, reassigned me to Cambodia at the end of my leave.

I was pleased that Monique, my bride, and I would have a normal life, since Cambodia was not at war. At the same time, I was disappointed because I had to leave Vietnam, my friends, and my personnel without bidding them farewell. Still, as I am not a masochistic man, I decided to look only on the bright side of things.

Monique and I arrived in Cambodia's capital, Phnom Penh, at the beginning of 1964. We were married there. I was honored with the presence of my long-time friend Dr. Paul Grauwin, author of *I Was a Doctor at Dien Bien Phu*, whom I had met for the first time in 1951 when he was a doctor in my battalion in Nam Dinh, North Vietnam.

A Cambodian plantation functioned as a Vietnamese one did; only the human factor changed. I found the Cambodians easy to live with. They were friendly, honest, frank, and not as calculating as the Vietnamese. Singing, dancing, carefree fun with family and friends, and an incapacity for work at 4:00 A.M. distinguished the Cambodian from his more austere neighbor, the Vietnamese. Accepting that no situation is perfect, we soon got on extremely well. I let them do what they wanted after work. Monique and I even participated sometimes, on condition that, whatever else happened, the work had to be done accurately and within the required time limit.

I was honored twice to dine with Cambodian leader Prince Sihanouk in his enchanting palace. A mercurial man, he was loved and respected by his people. He succeeded in maintaining Cambodia's neutrality for ten years while the communists in North Vietnam and the non-communists in South Vietnam fought their war along and across Cambodia's eastern and southern borders. Of course, this explanation is the official version of the Prince's policy. In reality, he closed his eyes to communist supplies and troops traveling along forest trails on the Cambodian side of the border.

This situation lasted until the end of 1969 when Prince Sihanouk took a private trip to France. General Lon Nol, who was deeply anti-communist, took advantage of this absence and, with some assistance from the United States, overthrew Sihanouk and established himself as the Cambodian Nationalist leader in March 1970.

A few weeks later, the Khmer Rouge emerged from their lairs. The Khmer Rouge—a name derived from Khmer, the Cambodian word for Cambodian, and Rouge, the French word for red—were assisted and accompanied by their communist associates, the North Vietnamese, who were more than pleased at any opportunity to get closer to their adversaries, the South Vietnamese and the United States. In retaliation, the United States, who had significantly reinforced their military presence in South Vietnam since 1964, made incursions and raids against Khmer Rouge and Viet Cong sanctuaries along the border.

The Cambodian Nationalists, who had lost their fighting edge during the ten-year period of peace, lost northern Cambodia in only one month to seasoned North Vietnamese working in concert with the Khmer Rouge and the Viet Cong. Chamcar-Loeu, the plantation that I was directing, fell behind their perimeter of "liberation." Its level of rubber production during this period was at its peak. Many of the hybrids in Cambodia produced twice as much per square mile as the trees in Vietnam had.

The communism that I had left behind in Vietnam had tracked me down in Cambodia, ending the ideal, trouble-free conditions that had existed there from 1964 to 1969. Gone were the days of visiting Angkor Wat, tiger hunting while riding elephants, and fishing on the Mekong River (among the most bountiful fishing water in the world). What had begun as an insignificant political movement had suddenly seized the nation.

After Lon Nol took the government away from Prince Sihanouk, I listened closely to both of my radios, the local one and my company's network, to try to keep abreast of the movements of both the Khmer Rouge and the Vietnamese communists. However, once again, the only dependable means of communication was the bamboo phone.

There were ten French families living in this area on two plantations: seven at Chamcar-Andong and three, including mine, at Chamcar-Loeu. Located in Kompong Cham province, the two plantations were about forty miles apart, which really could matter as far as communications were concerned. We were, of course, wishing for a miracle to stop the communist invasion, but none occurred.

When the bamboo phone at last alerted us that the communists were within one hundred miles of us, our general director asked us by radio if we would volunteer to stay at our posts. With the same attitude that I had had in Vietnam, I decided to stay. Although my wife wanted to stay with me, I convinced her that she and our six-year-old daughter should be evacuated to Phnom Penh, about one hundred miles from the plantation, where she could join her older sister and live with her family. This decision was all the more reasonable because Monique was pregnant. Her departure was heartbreaking but, as we knew absolutely nothing about what was going to happen, I was relieved to know that my family was safe.

Two Frenchmen remained on Chamcar-Andong plantation; I was the only Frenchman at Chamcar-Loeu. The dreaded day of the communists' appearance approached.

Eight

The Communists In Cambodia

The days following Monique's departure to Phnom Penh seemed to pass slowly. Living in my empty home, I tried to keep calm in spite of the anxiety creeping over me. The bamboo phone reported the Reds to be a few miles from us. One morning, when I was supervising an extensive area of young rubber trees about six miles from the plantation center, I saw them approach.

The small size of the young trees allowed me to identify them from several hundred yards away. The North Vietnamese, in their usual green khaki uniforms and tropical helmets, walked in a line on one side of the track; the Khmer Rouge, in black uniforms with red and white checked scarves around their necks, marched parallel to them along the other side of the track. This form of separate progress drew my attention. While I did not have time to make a detailed analysis, I stored the event in my memory —maybe it would prove useful. I already knew that the Cambodian and Vietnamese had not liked each other very much for ages. Even though they were all communists, their supposed "common" ideology may not have been able to bring them together.

As I was the only white man in the area, and as I was dressed in white shorts and shirt, I was easily spotted. The column stopped beside me and the North Vietnamese officer began to question me concerning the number of inhabitants and workers on the plantation and asked how many were foreigners. I briefly gave him this information, which did not commit me to anything. There were about eight thousand people spread out over three villages on the plantation. About one thousand of these were workers, three hundred of whom were Malaysian. I added that I was the only Frenchman and that I was responsible for this plantation. The word "director" was not in the communist vocabulary. The officer abruptly told me that I had to take all his soldiers to the central village and feed them. I cast a glance around me and I estimated the number of soldiers at approximately two hundred. One second later, I realized that the only available vehicle was my jeep! I told the officer that I did not have any trucks close by, but he did not care; quite to the contrary, he preferred a small vehicle because of the possibility of U.S. flights over the area. I had not thought about that and was delighted because it meant that they had already been pounded somewhere else. However, as I was, at the time, physically with them, and as the planes could not distinguish between us, the danger was real.

I calculated that at the rate of eight Bo Doi (communist soldiers) with weapons and kits per trip, I would have to make about twenty-five round trips of twelve miles each before I could end this boring and hazardous ordeal. Our tracks were bulldozed dirt levees and it was impossible to drive speedily on them, so I was anxious about conveying these troops before nightfall.

Not having been given a choice, I was forced to become a taxi driver and, as the first eight Bo Doi got into my jeep, my Cambodian overseer, Pok Chhuon, came close and asked in his language what was going to happen to me if I was taken prisoner. He wondered if he would ever see me again and what he should do. I put his mind at ease by explaining the situation and then asked him to talk to the workers (who had stopped working when the communists had arrived) and ask them to begin working again as usual. I then got into the jeep.

The North Vietnamese officer became suspicious at once and demanded to know exactly what I had said. He called a Khmer Rouge soldier, who had been resting on the ground, and ordered him to ask the supervisor, in Cambodian, what had been said. Pok

Chhuon, understandably afraid, confirmed that our conversation did not have any anti-communist or anti-revolutionary purpose and that I had only wanted to put their minds at ease. The officer seemed satisfied with this answer, but he favored me with a long boring sermon on obedience and communism. It was as if I had been transported back six years to the time of my Vietnamese nightmare. Furthermore, as our invaders appeared to be extremely suspicious, I felt that life in the coming weeks and months was not going to be easy. After losing an hour for nothing, I moved off with the first group of Bo Doi holding on to the jeep. I watched the sky while I drove because the rumbling of my engine might drown the sound of U.S. planes and I did not feel like being shot.

This first trip was rather hard. The Bo Doi wondered loudly if I was a spy, if I would drive them to an unforeseen place, if I kept in touch by radio with the U.S. planes, and so on. I asked them to stop talking such rubbish for, if something serious should happen, I was among them and as vulnerable as they. My comment seemed to calm them down so quickly that I at once understood that their apparent concern was only intended to make me feel insecure. I decided to drive on silently, maintaining my watchful eye on the sky. As the bumpy levee and my unbelievable load made the driving dangerous, I was unable to make more than ten or fifteen miles per hour.

Our arrival in the central village prompted an alarmed gathering. When the inhabitants saw me surrounded by a pack of foreign soldiers, they thought I was a prisoner. In fact, I was not a prisoner in a jeep or in a hut, but I was, in reality, a prisoner inside the invaded area. I explained the situation to them and gave them some news about their relatives, who had been with me when the communists arrived.

I dropped the Bo Doi under the adult rubber trees along the edge of the village in order to avoid a possible attack against village families by U.S. planes. If there was to be a bombing, I preferred it to be in the forest. The North Vietnamese officer ordered one of his Bo Doi to ride with me when I returned to pick up another cluster of soldiers. Obviously, he did not trust me!

I would have been able to call a driver with a large truck to make the round trips, but the situation was dangerous enough as it was. I was afraid of a possible negative reaction by a Cambodian driver not trained to handle this type of situation, so I decided to continue this unusual mode of transport alone. I explained my decision to the North Vietnamese officer, but he made no comment.

From 9:00 A.M. until 6:00 P.M., I transported communists. I became completely fagged out—from both the physical and the psychological strain. Visions of my family and bombings kept my entrails in knots and my mind in bewilderment. A few days prior, U.S. planes had hit various locations on my plantation each day from a low altitude, but each time, by coincidence, I had been either under the adult rubber trees or in the plant or in my office.

That day, as I ferried troops, it was different, but God, no doubt, was still with me. After about a dozen trips with a guard, the North Vietnamese officer allowed me to drive alone on the return trips as the afternoon progressed. On one of those returns, a U.S. reconnaissance aircraft flew so low over me that I could clearly see the pilot make friendly gestures to me several times. Being sighted by that pilot was a close call, and I was worried sick over that incident for several days.

When this infernal public transport ended, although I was worn out, I had to meet my staff and personnel in order to reassure them and specify that we had to continue to work as normally as possible. The Malaysian employees were especially difficult to please. They wanted to go back to their country and get away from the war. The North Vietnamese officer, hearing us talking loudly, asked why the workers and their families screamed like that. After having been the driver, I now had to be the interpreter. That was the last straw! Nonetheless, I explained to him the Malaysians' frame of mind. He requested that I tell them that his troops of liberation would not tolerate any disorder. His Bo Doi were ordered to fire on any people who might want to run away. I refused to relay this threat to my employees, so the officer gave me a hard time for a few minutes. He even drew his handgun and brandished it in front of the crowd. The situation changed quickly when the Malaysian crowd, who refused to be involved with the war, began to scream and move forward. The Bo Doi ran up to us, slamming closed the breeches of their guns, and the North Vietnamese officer fired into the air several times. There was an immediate and complete silence.

The Khmer Rouge leader, who was with his men not far from where we were and who had not participated at all in the meeting, approached the North Vietnamese officer and told him, in Vietnamese and in a very low voice, "In this kind of situation, I never discuss or explain. I order and, if I am not obeyed, I fire." Both of them were close enough for me to understand them and my impression was that the Khmer Rouge, whom nobody had really known until this moment, were extremely dangerous. Unfortunately, this impression was verified over the following weeks. At the time, I quickly made a discreet sign to my employees to go home at once, hoping to stop the baiting of this trap. Fortunately, the maneuver worked and I could breathe again.

I noticed that the women of the village, asked to cook for the Bo Doi and the Khmer Rouge, killed the old fowls (keeping the young ones for us). In addition, one of my mechanics caught a fifteen-foot-long python, which was skinned and cut into slices. Some of the Bo Doi, who were fond of dog meat, caught three or four stray dogs that they brought for the women to cook. Ordered to do so by their parents, many of the children caught fish and frogs in the surrounding lakes and rice fields. All these provisions enabled us to feed the more than two hundred soldiers, but what would we do after that?

This strange meal, with its various and unusual aromas, reminded me of a commonly held, local belief: when a group of dog eaters patrol a village, all of the dogs will bay; when non-dog eaters patrol, the dogs will not. This rumor proved to be true, as I was to witness. I was dumbfounded that the dogs' sense of smell was so keen they could detect the stench of death from the living human beings.

My staff invited me to dinner, more to talk than to eat. Everybody was worried about the future and was struggling to find some sense of security. The problem was that I also felt insecure. It was not easy to put their minds at ease when I was gripped by the same fears; nevertheless, I tried, and, after a speech that contained comments on peace, war, rubber production, friendship, python (which I liked better than dog), and old fowls, I finally was able to go home and flop down on my bed.

Despite being exhausted, I rose at 4:00 A.M. and, in order to show everybody that nothing about work had changed, I drove through all of the alternate bleeding areas of the day. I decided to try to meet with the North Vietnamese officer, whom I considered more approachable that the Khmer Rouge leader, to ask him to move his troops off the plantation. I knew this was a risky step; however, the risk was better than having my three villages and my plant bombed.

I easily found the North Vietnamese officer, to whom I explained my point of view. I had transported and fed them all the day before, as they required of us. They even had slept in one of our sheds. However, they had to leave the plantation as soon as possible. He asked me why I wished them to leave and, as usual, he began a long speech about the validity of the Khmer Rouge's revolution in particular and about the future communist hegemony in general. He was there himself with his troops only in order to help his Khmer Rouge comrades in the liberation of their country. Paradoxically, I found myself using the same arguments I had with the Viet Cong in the early sixties: for example, several thousand people lived on and depended upon the plantation; if it were bombed and destroyed, all these people would lose their homes and their incomes; this possible scenario could impact his own situation. The only difference between the Viet Cong in Vietnam and the North Vietnamese in Cambodia was that the North Vietnamese were not in their own country. The Vietnamese officer listened to my speech and told me he would have to consult with the Khmer Rouge leader about my request.

That evening, all the North Vietnamese and Khmer Rouge moved to the forest, about two or three miles from us, and settled into an old pagoda that was no longer used for worship. I learned a few days later, through one of my workers who had discovered a remote cousin among the Khmer Rouge, that the Khmer Rouge troops were very averse about this decision and were saying that they would kill me if I didn't stay in line. When I received this information, I was in a cold sweat, and I swore to myself that I would stop doing such risky things. I had to learn to adapt my behavior and be more of a family man, especially with such raving madmen as the Khmer Rouge. I thanked God one more time for having inspired me to choose the North Vietnamese to consult about my desire to have them out of my way.

The following days, the bamboo phone brought news about the Khmer Rouge's intransigence. They were trigger-happy and never had any friends among the villages they occupied. Of course, for the people, being neutral was not a choice. They either had to support the communist cause or die. This news neatly matched my own experience and confirmed my first impressions: I was going to have to be very careful.

In Vietnam I had gained some experience about how to manage many difficult situations, but this experience was almost valueless in Cambodia. Indeed, in Vietnam, I had been on a plantation located in a part of the country where communist infiltrators were almost everywhere, even though the official power belonged to the non-communists helped by their U.S. allies. One important advantage to that situation was that we were able to rely on all the commercial, trading, administrative, and banking structures to work as usual. Everything was the opposite in Cambodia. The situation on the ground was much clearer and, paradoxically, much more difficult. We were located in a part of the country where the power belonged to the communists and the business structure was nonexistent. This happened because a temporary halt in the progress of the communist troops had split Cambodia and, in the process, we were cut off. Moreover, as the North Vietnamese and the Khmer Rouge did not have the same methods of indoctrination or culture, there were many areas in which they did not agree. Over the next several months, I found it necessary to use this tension to antagonize these two occupational forces against one another in order to protect the plantation and its workers.

With the passing of time, I cannot help but think of how the traditional contempt between the Vietnamese and the Cambodians has played itself out in this region. When the Khmer Rouge invaded Cambodia in 1970, the North Vietnamese, who at that time were engaged in a war against South Vietnam and the U.S., opportunistically helped the Khmer Rouge in their fight against General Lon Nol, an ally of the United States. In 1975, when Saigon and Phnom Penh fell to the communists, the Khmer Rouge seized power in Cambodia and indulged in one of history's bloodiest holocausts. In 1979, communists from Vietnam *invaded* Cambodia to wage war against their former ally and stop the holocaust.

No one will ever know if the Vietnamese communists invaded because of their concern over the plight of the Cambodians or because of their concern to show the world that the "true" communism was theirs. At the risk of disappointing those who think that compassion might have prevailed over ideology, I personally believe the second option to be the right one.

Nine

Daily Life in the Communist Zone

Keeping the plantation in working order was about to become a *tour de force*. We were dependent on transportation for all of the direct and indirect ingredients essential to rubber production. From the Mekong River and the jungle, we managed to sustain ourselves with meat, fish, and rice. However, the artificial border, resulting from the temporary halt of the communist invasion, became so compartmentalized that I had to find some courageous Cambodian faithmen—dedicated, trusted men—to recruit smugglers for gasoline, diesel, paraffin, formic and acetic acids, all specific rubber ingredients, and the hardest but most essential to find, money.

Most roads, especially those to the south, the sea, and the capital, were cut in "piano keys"—a series of overlapping trenches dug along each side of the road that blocked all four-wheel vehicle traffic—a method familiar to the French from the previous war. This obstacle made communication with Phnom Penh impossible since the Khmer Rougė had shot a hail of bullets into my radio the first day they arrived to keep me from talking with the "traitors" in the capital. Fortunately, my two colleagues at Chamcar-Andong had been luckier than I, their radio communication worked as it should. Perhaps they had encountered a more understanding and resourceful communist visitor. In any case, they were forty miles away and the roads were neither easy nor safe. I bought a black-market motorcycle to use when traveling and used the jeep only on the plantation.

A few days after the unwelcome communist invasion, during which time we worked with limited supplies, transport got as organized as possible and rice, as well as gasoline, arrived by sampans, junks, bicycles, motorcycle trailers, and ox-drawn wagons. Some supplies, such as acid and ammonia, were carried in by hand. Everything else we needed reached the plantation haphazardly and at any time. Occasionally, nothing arrived. Generally, smugglers delivered goods twice a day. As all supplies had to cross the fictitious mid-country border, they arrived in scattered disorder. We grouped items together for storage in our warehouse and, after I had okayed everything, the smugglers were paid in cash. Perversely, I could not do the same with our rubber product. Ten tons per day had to travel in a unit with four faithmen in order to avoid possible misappropriation or theft as well as to guarantee a cash return without which we could not continue to work and live.

With no radio or phone contact possible, Chamcar-Andong and Chamcar-Loeu corresponded by bicycle or motorcycle messenger. As the need to continue the rubber shipments was an emergency situation, my two colleagues from Chamcar-Andong (who needed to ship fifteen tons per day) and I decided to meet with the North Vietnamese commanding officer of our area in order to get a concession for this vital transport. Although I had persuaded my two friends that it was better to be faced with the North Vietnamese only, two Khmer Rouge leaders were also present at our meeting. I was disappointed, but I realized that the North Vietnamese doubtless had to include their Cambodian allies. In spite of the unwanted Khmer Rouge being present, we explained that if the rubber could not be carried south and sold there, we would not be able to pay the personnel. That would have serious consequences on the stability of the close to twenty-five thousand people living on the plantations and, consequently, on their own stability. We saw that the North Vietnamese understood the situation, but they could not soften their communist allies. The Khmer Rouge did not want to give "their" rubber to the traitors in Phnom Penh.

Time passed in fruitless discussion, each party reiterating his own point of view. We got nowhere. After a while, the North Vietnamese and Khmer Rouge leaders went outside to talk. When they came back, the stubborn Cambodian sounded a little less stupid but just as narrow minded as before. For example, in order to transport twenty-five tons of rubber a day, the piano-key trenches along the road had to be filled. The Khmer Rouge said we could fill in the trenches as long as we re-dug them as soon as the trucks had passed. This condition was completely irrational and impossible to sustain over a long period of time. Even though we tried for several hours to explain, we obtained no concrete concessions. The North Vietnamese, who seemed bothered, left the room first; we immediately followed because it was out of the question for us to remain alone with the Khmer Rouge.

My two friends from Chamcar-Andong and I decided to try to comply with the Khmer Rouge's instructions. We felt that, had we not obeyed, they would have closed the plantations and arrested us as traitors. I appointed three hundred people, as did Chamcar-Andong, to fill the trenches along the road to the south for about twenty miles. This had to be done at night so that no reconnaissance aircraft could alert the bombers. We held out for three days and nights, doing our usual work as well. Finally, after wasting our energy on this moronic and impossible task, we abandoned it.

I was, in fact, in the middle of a delicate balancing act. I had to keep the plantation functioning as well as could be expected. I also had to take into consideration the local population's total dependence upon the plantation as well as the communists' partial and subtle interdependence with the local population. And, I had to manage to get the rubber harvest to market without anyone losing face.

In addition to the absurdities described above, other difficulties cropped up—such as nondebatable daily requisitions from the "Liberation Army" that required one's instant participation. After a few days, my two Cambodian assistants had to abandon their jeeps to the communists, and the plantation had to continue to provide both gasoline and maintenance. My assistants had many problems getting around the plantation to supervise the different areas. They managed on their own by borrowing my jeep—which the occupying forces had, for now, left to me—or by using the trucks that carried both workers and latex, or by riding a bicycle. After about a month, the Khmer Rouge told me that it was intolerable to see me driving a jeep while they were walking. Ultimately I had to give it to them, thereby furnishing them with three vehicles we had to maintain.

It was impossible for me to do my job correctly over such a large area without a vehicle, especially with more and more problems developing both all over the plantation and outside of it. Consequently, I used my Honda motorcycle every day. This solution would have been excellent if I had asphalt on which to ride, but this was not the case. All the levee trails were hard during the dry season; they looked like corrugated iron. During the rainy season, they were as slippery as a skating rink. I took an average of two tumbles a week during the six months of the rainy season, fortunately without any grave results; the soaked red soil cushioned my landings. The motorcycle had only one advantage; I could dodge in and out of the piano keys on the Chamcar-Andong road. Even this advantage was of limited importance because, when possible, I avoided taking the roads. Anything —being shot, arrested, savagely questioned, or worse—could and did happen to people traveling along the roads. Earlier, Sean Flynn, son of the famous actor Errol Flynn, had disappeared without a trace along a road not far from our area.

After a few weeks of this unruly and tiring life, I felt stifled. I needed to speak French! I decided to make the trip to Chamcar-Andong, even though there was no guarantee that I would be able to return. Before I made this decision, I thought for a long time about Monique, my wife, and Dominique, my six-year-old daughter, both wilting in Phnom Penh. We had been communicating over the past month via short notes carried by an obliging smuggler. I considered Monique's pregnancy and anticipated the happiness of our having another child. I wondered if it was reasonable to travel this once-familiar road that had become a dangerous unknown due to the uncertainties of war. Finally, my spirit of adventure and my forced loneliness combined to urge me to go.

Because I anticipated that I would have some problems, I took the precaution of starting in the middle of the afternoon in order to arrive before nightfall. On and around the plantation, the North Vietnamese and Khmer Rouge knew me, or at least had already seen me; on the road it would be different. I asked the North Vietnamese leader for a pass, thinking that it would help me along the road. I was too optimistic. After a few miles, I was stopped by the first Khmer Rouge roadblock. One after another they examined my pass, asking many aggressive questions. Although I was in a blue funk, I gave them explanations that obviously they did not want to understand. After a long discussion among themselves, during which I heard the word "spy" repeated several times, I was instructed to go back to the plantation and get another pass—a Khmer Rouge pass. While my North Vietnamese pass had not allowed me to go through the roadblock, I congratulated myself because it had prevented me from being arrested.

I returned to my plantation and explained to the North Vietnamese officer that I had to meet with my colleagues at Chamcar-Andong to discuss certain techniques related to rubber production. He seemed rather irritated when I asked him to go with me to meet the Khmer Rouge leader. I preferred this approach over going directly to the Khmer Rouge leader, which I felt would have gained me nothing.

I eventually got my pass and, even though I had lost about two hours, decided to attempt to make my journey again. The Khmer Rouge at the first roadblock recognized me and let me go after they checked my pass. On my way, I passed one Bo Doi company a couple of times; some of them made signs, shouted out, and smiled at me. Obviously, they took me for a Russian advisor (at the time, the North Vietnamese were being aided by the Russians and the Khmer Rouge were being aided by the Chinese). For them, a white man, freely moving about in that zone, could not be anyone else.

The second roadblock was North Vietnamese, and I had no problem going through with my pass. The next roadblock was Khmer Rouge and there I had some problems explaining why I rode a motorcycle and why I had a communist pass although I was white and apparently a non-communist. They asked and I had to respond. Finally, one of them, no doubt not quite as short-sighted as the others, let me go when he heard an aircraft flying in our direction, giving me only enough time to rush face down under some trees with my motorcycle before the plane shot gusts of machine-gun fire on the road. Continuing my journey a few minutes later, I was again in a blue funk until I reached the next North Vietnamese roadblock. The leader there was impressed with my pass and let me continue at once. At nightfall, after taking six hours to travel only forty miles, I arrived at Chamcar-Andong. What a relief it was to meet two nice, friendly guys to whom I could tell my misfortunes while drinking some scotch; I forget how much. We commiserated about our losses, such as my rifle, to which I had become sentimentally attached. Of course, the greatest loss was our freedom.

The tale concerning the loss of my rifle is a long one. In the first days following their arrival, the communists had demanded that everyone surrender their arms immediately or face death. Two days later, none of the workers and their families had given up anything. I was not surprised. The people who lived in my three villages were calm, quiet, and relaxed; most of them had never seen a gun. In fact, all of the people, including myself, had interpreted the word "arms" as "guns." However, the communists, especially the Khmer Rouge, defined "arms" as "anything that could serve as a weapon."

The Khmer Rouge took a dim view of everyone's failure to turn in weapons, so they rummaged through every house. I suddenly thought of my rifle at home. Somehow, I had failed to realize that the order to hand over weapons also applied to me. I guess the emotional baggage caused by the communists' arrival had caused me to overlook this detail. Maybe because I saw myself as being there only to produce rubber, I failed to realize that my hunting rifle could be considered a threat. Their search progressed and the so-called arms piled up in the center of each village: machetes, hunting knives, hunting spears, and one handgun. A poor Montagnard worker refused to give up the spear that his father had given him, which he used only for hunting, and was shot down on the spot by the Khmer Rouge who were maddened, on the one hand, by the hapless man's resistance and, on the other, by having already found a handgun.

I immediately sent a member of my staff to talk to the North Vietnamese officer about this horrible and gratuitous murder. By the time he and some of his men got to the scene, a crowd was brewing and I wondered when the Khmer Rouge would reach the stage where they would shoot people. The Bo Doi fanned out and, to allay any mob hysteria, shot into the air. This surprised everyone, including the Khmer Rouge, and restored order.

The Khmer Rouge and North Vietnamese officers engaged in an extended consultation, which was followed by the North Vietnamese leader giving a speech, translated by a Vietnamese Bo Doi from Cambodia. Once everyone had calmed down, some of the workers explained that they could do nothing without machetes, axes, and hunting knives, and that they did not look on these items as arms but as essential tools. The two communist leaders consulted one more time and, after a few minutes, the Khmer Rouge left the tools but arrested Por Try, one of my three Cambodian assistants, because they had found the handgun in his house.

After everybody had retrieved their tools and gone home, the Khmer Rouge tied Por Try's hands and took him away—but not before he glanced despairingly at me. I knew he was in a bad mess, and I wondered how I would be able to help him.

At this moment, I realized that the Khmer Rouge had not yet rummaged my home, located about three miles away. I never knew whether or not this oversight on their part was involuntary, but I accepted it as a blessing. As I could not run the risk of my rifle being discovered by the Khmer Rouge at some future point in time, either through betrayal by a dismissed worker or search or some other occurrence, I decided I had to dismantle it and remove it from my house. My heart sank at the thought; my American-made M-1 was a war souvenir that had saved my life several times in North Vietnam. Anyway, I had no choice. If the Khmer Rouge found it in my possession, they might think themselves justified in taking reprisals against my family in Phnom Penh. My family's safety was far more important than the company's rubber. I broke the M-1 down, took one piece at a time, and threw each piece in the jungle or in the river where it would never be found. This plan worked smoothly with the exception of one dicey moment when I unexpectedly had to act as driver for four Khmer Rouge while I had the breech of the gun hidden in my briefs. That incident left me sweating!

As soon as I had disposed of my rifle, I tried to see what I could do for Por Try.

Ten

The Arrest of Por Try

I called together my two remaining assistants, Nou Try and Sok Chuon, and some others who were especially smart and reliable. All of us knew Por Try as a good, honest man, and everyone agreed we should do something to attempt to rescue him, but nobody really had any idea about how we should go about doing this. In fact, each man was afraid. Slowly, one by one, every head turned to me. We all, including myself, feared the Khmer Rouge, and with good reason. Unanimously and simultaneously, the thought that a Frenchman who was the director of the plantation had a greater chance of succeeding than they took hold. Objectively, I agreed; deep inside, I wondered what kind of mess I was getting into this time.

We decided to start by talking to his wife, who was distressed and shaken. We learned that her brother had given the offending handgun to Por Try a few years before as a gift, that he knew nothing at all about guns, and that the weapon had never left their home. It was a very weak angle. However, the importance attached to gift giving in Asia was a strongly shared value that cut across cultural boundaries. Perhaps I could play on that concept.

Through a worker who was a little more collaborative, motivated, and opportunistic than the others, we managed to learn that Por Try had been taken to a small Khmer Rouge headquarters in a village located at the edge of the jungle, about twenty miles from the plantation.

That unwelcome news meant that I had to play with fire one more time. I had to travel alone, by motorcycle, through a zone of instability swarming with Khmer Rouge in order to rescue a victim arrested by an uncontrollable communist unit. It was a project of the wildest extravagance, but I had to do it anyway. Also, I had to act rapidly if I wanted to preempt the swift justice of a Revolutionary Court. No doubt this court would be even more expeditious than the People's Courts in Vietnam, and Por Try would stand little chance.

Before starting my journey, I took two principal precautions. First, through the worker who had directed me to Por Try, I got a written pass so that I could move into the area without an exact destination. Second, I requested Nou Try to alert the two Frenchmen at Chamcar-Andong if I was not back after three days. After that, I set off on my dangerous adventure.

As usual, the Vietnamese did not stop me except at the roadblock, where they contented themselves with making half-hearted gestures in my direction. My dreaded first contact with the Khmer Rouge occurred a little further along the forest tract when I ran into one of their patrols. They surrounded me at once, making a great clamor with their shouts and pointing their guns at me. My pass made them suddenly incredulous, circumspect, and suspicious.

As I had to give some information about my presence there, and as I could not tell them the truth, I told them I was searching for a lost friend. If I had been honest and told them that I was searching for someone who had been arrested by their revolutionary comrades, they would have stopped my trip. After a long discussion, they hit upon the idea of designating one of their number to accompany me to the next Khmer Rouge control point. Without any further explanation or comment, they ordered me to leave. Uncomfortable, I wondered if I was free or not, and I feared that my half-truth might work against me when I was at last on the trail of Por Try.

Anyway, I let out the clutch of my borrowed Enduro Honda and went on, with the Khmer Rouge seated on the saddle behind me. I was amazed at his reaction; he was smiling and seemed to be having a ball. I had never seen a Khmer Rouge smile, and he would be the last smiling one I did see for a long time. This being his first ride, he sat stiffly, not flowing with the movements of the motorcycle. I took advantage of his inexperience by taking risks and riding as fast as I could; a small consolation.

My guide, my guard as well, pointed out the very small village in the forest where the next Khmer Rouge control point was located. Our arrival created a sensation. I had no doubt that most of the people there had never seen a motorcycle, much less a white man and a Khmer Rouge traveling together. White men were symbolic of the hated Americans. A revolutionary-communist-totalitarian comrade riding a motorcycle with a supposed American capitalist provided the spectacle of a lifetime!

My passenger gave the necessary explanations to the other Khmer Rouge and walked away. By chance, he said little about my objective, so I asked them the way to the village indicated by my worker. They checked my pass, showed me the right direction, and let me go.

I reached the village in question without any major problem except that two monkeys fighting on the track knocked me down, fortunately without doing any damage to me or to the motorcycle. I wondered if my informant had been right, as I did not see any Khmer Rouge. I began to ask questions of the inhabitants who, though surprised to see a Frenchman in the area at such a time (especially one who spoke their language), were nonetheless friendly. I introduced myself and engaged them in conversation during which I learned that they knew nothing of a man arrested by the Khmer Rouge. Finally, I did learn that the Khmer Rouge camp I was looking for was just north of this village.

My first contact was very disconcerting and distressing. As soon as the Khmer Rouge saw me, they took my motorcycle and tied my hands behind my back before I could show them my pass. And these were the people I had to talk to, if I could, about Por Try.

Contrary to my expectation, when I was finally able to show my pass, it caused them to take me for a spy. That was the last straw! The pass, which was supposed to help me, plunged me deeper into trouble. I was both worried and angered as I tried to explain the actual situation: who I was, what I did on the plantation, and who I knew among their communist representatives, both North Vietnamese and Khmer Rouge alike. I continued with how I got the pass, adding that if I failed to return to the plantation rapidly, there would be no work for the plantation hands, no income for them and their families, and more problems for the troops stationed in the area. This hard discussion, conducted with my hands tied behind me, dragged on for three hours before one of them, perhaps a leader, ordered four others to take me out and lock me up in a hut located a little farther away.

I stayed there two days and one night, deeply depressed. I even came to miss my stays in the hut in Vietnam where I had been bored by the many political speeches each day. At least those had kept me busy. Among the Khmer Rouge, I received no word and no one visited. Nothing happened other than my being given a little rice twice a day.

At the end of the second day, I was allowed to go out of the hut and meet the same Khmer Rouge I had seen the first day. He seemed less stubborn and told me abruptly that, contrary to what he had believed two days earlier, he now believed that I was not a liar. During our conversation, he told me that they had sent an emissary to the plantation in order to verify my story and inform my people of my plight. Compared to my journey thus far, the task at hand, Por Try's release, was monumental. It appeared the Khmer Rouge were willing to allow me to continue my search.

Wondering what to do next in order to locate Por Try and negotiate his release, I came up with a harebrained idea. Remembering how my Khmer Rouge guide/guard had flipped over his motorcycle ride, I thought maybe my two hut guards would react the same way. Just before leaving, with the engine running, I asked them if they wanted a ride. They seemed more young than hardened, and they accepted my proposition. As they had been ordered to stay together, they insisted we ride three at a time—weapons and all. I still managed to build up some speed and give them a bit of a motorcross experience around curves and up and down hills—even managing to include one good fall. The whole exercise opened up conversation and, hopefully, some level of trust, as we shared impressions and experiences. Still, I knew that anything I had gone through up to now was minimal to what lay ahead. What a job!

I was jubilant when my efforts succeeded! I learned that there was a prisoner at this camp who was accused of betrayal, espionage, and the illegal carrying of a firearm. As every step and word I took was being watched and weighed, I decided to wait long enough for my two passengers to have time to tell the others about their ride before meeting with the local Khmer Rouge leader.

After a while, I threw myself into the fray and went to meet the Khmer Rouge leader. With fear gripping my throat, I heard myself asking for some news about Por Try. The leader did not seem to be in a negative frame of mind; hopefully he had heard of the motorcycle rides. He asked me why I needed information about a traitor, so I told him in detail who Por Try was, what his job was, and how he generally behaved on the job. I went on to explain that I had learned that the handgun had been a wedding gift, and that Por Try had never used it. Then I asked permission to see him. Much to my surprise, he took me to meet Por Try, who was tied up in another hut. As soon as I caught a glimpse of him, I began to bawl him out, and continued to give him a virulent chewing out for several minutes—and I did it in Cambodian so everyone would understand. As I went on, I nervously wondered what the Khmer Rouge's reaction would be.

They seemed dumbfounded. Obviously, they had not expected this outburst from me. Having finished with Por Try, I seized the opportunity and, telling the Khmer Rouge that I would deal with Por Try's errant behavior, I asked them to turn him over to my custody. They agreed! I have never understood how the

Khmer Rouge came to be so agreeable that day since they could have sent a bullet through the two of us at any point in this charade. Not waiting for them to have second thoughts, Por Try and I—hungry, filthy, thirsty, and tired as we were—immediately left on the Honda, hoping that there was enough gas left in the tank to get us home. Running out of gas was just the sort of silly detail that could have turned our triumphant ride home into a nightmare.

Having gotten a double pass from Por Try's jailer, we rode away down the forest track. Por Try told me that he had recovered his confidence as soon as he had seen me. His concept of faith did not seem to fit the danger of his situation; still, I was feeling quite proud of my rescue effort even though I was in a cold sweat.

Fortunately, our ride back went smoothly and, as soon as we entered the plantation's central village, a howling mob gathered around the motorcycle, congratulating and kissing both of us. A big celebratory meal was held, during which *choum*, a rice alcohol, flowed freely. By 4:00 A.M. the next morning, few people were not seeing double, myself included. However, everyone made it a point of honor to be up and on the job, never mind if the bleeding notches on the rubber trees were a bit twisted. During such exceptional moments, I was acutely aware of the value of the mutual friendship and confidence that the people of the plantation and I had established. We had gone through a difficult time together during which our deep and common bond had supported us all.

This type of adventure, peppered with similar ups and downs, happened again about three months later when two other assistants, Sok Chuon and Van Doi (a Sino-Khmer-Vietnamese supervisor), were arrested by the Khmer Rouge and charged with spying and having contact with the enemy. That time I had to convince the communists that Sok Chuon and Van Doi had acted according to my orders, and that they really had tried to get in touch with some peasants from the next province located in the "non-liberated" zone in order to negotiate rice purchases for the plantation. I implied that if the plantation could not eat, the troops could not either. This argument worked, but we had taken a serious risk because Sok Chuon had been supporting his family and Nationalist Army friends in Phnom Penh, where he had been born. In addition, I learned that Sok Chuon had been giving information about the Khmer Rouge to Nationalist sympathizers. A few days later, Sok Chuon—an intelligent man who understood the situation perfectly and who did not want to endanger either the plantation, his friends, or me—escaped from the Khmer Rouge zone during the night.

Eleven

Air Raids

Because my plantation was located only a few miles from the artificial line that separated the "liberated zone" from the rest of Cambodia, and as there were United States Air Force bases not far across the border in South Vietnam, our zone was subject to constant surveillance. I understood perfectly that the pilots were trying to observe communist troop maneuvers. This, of course, was why I had asked the communist troops to camp away from our village. However, we dreaded the U.S. air raids more than living among the fickle Khmer Rouge. The Khmer Rouge dealt out death swiftly and unpredictably; bombs were definite and deadly. The Khmer Rouge could sometimes be influenced by words; bombs could not.

Through 1970, fortunately for us, we saw many more reconnaissance aircraft than fighters or bombers. But, if danger did not then rain from the air, it did come from the ground because the Khmer Rouge and North Vietnamese, although not camped on the plantation, swarmed over and patrolled it constantly. The communists, too, were fearful of the aircraft, and not only for the obvious reasons. They kept a watch on reconnaissance raids to prevent inhabitants from picking up tracts that sometimes were dropped over the area. There was also the threat of coded messages being exchanged between a worker, spying for the South, and a plane. Thus, in the cleared areas, for me to wave or make any other hand signal to an aircraft, or for an aircraft to dip a wing acknowledging my presence, was very risky and dangerous.

When supervising in an extensive open area, while dressed in my white clothing, I easily stood out from the plantation workers, who were dressed in their black and brown clothing. When an aircraft passed over, any communist troops in the area would scurry to the nearest stand of adult trees and hide. I did not know if the U.S. pilots had any information about the few Frenchmen scattered among the plantations, and I hated this situation in which I was forced to reluctantly conspire with the enemy by not indicating their whereabouts. However, as I valued my skin more than I resented the role, I endured it whether I liked it or not.

My greater worry was whether or not U.S. pilots could distinguish a worker dressed in black from a communist soldier also dressed in black. As the communist troops had some clever means of camouflaging their guns, the distinction would have been impossible from the air. I believe that this type of confusion resulted in the plantation being fired upon several times, usually along the tracks of rubber trees. However, for reasons I was unable to understand, my house was hit twice. Fortunately, I was at work when the attacks occurred, and my cook and my guard were unhurt. Once, shells loaded with grapeshot came right through the roof; three "secured" the bathtub! The damage was not very important on the overall scale of things, but it did frighten me. I had enough trouble coping with the communists. Air raids, especially against my own house, could be the end of me. Fortunately, unlike some other plantations located close to the South Vietnamese border that were almost destroyed by chemicals, we were lucky enough to avoid being sprayed by defoliant.

To make matters still worse, these attacks brought about yet another problem. Every plantation had its own grass landing strip that was used by the company plane on an average of once a week to transport the doctor who examined the patients in the plantation infirmary and, on occasion, to evacuate the sick or injured to Phnom Penh. These landing strips were maintained in good order but at a high cost because termite nests, which grew like mushrooms, could become as big as wheelbarrows if we did not take care of them. Also, wild banana trees grew rapidly, from one to three feet a week during the rainy season, and they had to be cleared. With the coming of the communist troops, I had stepped up my efforts to maintain my landing strip in the best condition possible, for I could not help thinking of it as my ultimate avenue of escape from the mess in which I was living.

The communists, who could no longer endure the thought of our landing strip being usable when U.S. aircraft were so close, ordered me to scatter and bury several hundred huge poles over the strip. My morale sank. Given no other choice, I began this difficult, stupid, and unremitting work with a heavy soul, working as slowly as possible. I felt we were cutting our umbilical cord as we destroyed what had taken so much effort to build and maintain. With the loss of this last outside link, I was now totally cut off from the civilized world. Even so, I had to continue to produce rubber, which I did as effectively as possible given the situation.

My company's other plantation, Chamcar-Andong, located forty miles further away from the pseudo-border, had been luckier than mine because it had been allowed to keep its radio and landing strip in working order. One day, about three months after the invasion, I received a short note by messenger from Chamcar-Andong saying that the general director was trying to hire a private pilot to fly in replacements so the three Frenchmen could go to Phnom Penh and decompress for a couple of weeks. We finally succeeded in getting the communists to agree to this project on condition that we stay in Phnom Penh only ten days.

This relief flight took place in May and, the moment the plane landed at Chamcar-Andong, the Khmer Rouge and North Vietnamese, who had been hidden under the rubber trees that lined the landing strip, surrounded the plane with their guns at the ready, intending to impress the two newcomers, the pilot, and the three of us. All of us were brought together to the edge of the field and put under guard while the communists searched both the plane and the luggage. The volunteer pilot, who had proven his merit by flying in but who had never before seen the communists at such close range, panicked and kept saying, “We have to escape. They are going to cut our throats!” He raged on and on until the three of us, who had survived these conditions for months, pointed out that (a) we had nowhere to run, and (b) the best way to get killed was to try to escape.

Of course, the communists did not find anything in the plane, but they discussed the matter with each other under the rubber trees for about two hours, no doubt in order to worry us and show us that they were in control of the situation. In fact, anything could have happened at this point. They could have kept us all hostage, or they could have eliminated us and kept the plane. Once again, we thought of Sean Flynn, who had been killed in our area by the Khmer Rouge. The situation was fraught with frightening possibilities but, fortunately, they kept their word and the exchange was carried out. As neither of the replacements knew Chamcar-Loeu, we agreed, before taking off, that both replacements would stay at Chamcar-Andong. Besides, the trip between the two plantations for a mere ten-day stay would have generated more problems that it would have solved. Moreover, the real reason for this exchange was to provide the communists with a few French hostages who could be responsible for the operation of the plantations. After all, the plantations provided sustenance to thousands of people living in the zone and, indirectly, to the Khmer Rouge and North Vietnamese as well.

My ten-day stay in Phnom Penh passed like a flash of lightning filled with joy. Monique and I thought that, if our time together could have lasted a month, life would have eased a little. The first days promised happiness, the recovery of psychological and physical strength, and the strengthening of moral fiber necessary for our precarious existence. But this process was all too quickly interrupted by the harsh reality of having to prepare to leave. Caught up in a system to which we had to yield, Monique, with her pregnancy, and I, sunk in my nightmares, were easy prey for stress. Understandably, there were many tears when I was about to leave. My little six-year-old girl did not understand why I was leaving and, although I tried to put my wife's mind at ease, I was merely feigning confidence.

The takeoff was a heartbreaking one; nobody knew if this trip was the first of many or the last of all. Personally, I was in a really bad predicament. I not only had to leave my wife and daughter, but I did not know if I would be able to be present when my next child would be born in August. Furthermore, I was angry because I had learned that our pilot had remarked in several restaurants and bars around Phnom Pehn that we would never be seen again and that "staying over there is suicide." If those comments got back to my wife, they might adversely affect her delivery.

The return trip, with another pilot, went off without a hitch, and our two substitutes took off at once. I retrieved my Khmer Rouge and North Vietnamese passes, which I had left at Chamcar-Andong to avoid their possible confiscation if searched at the Phnom Penh airport, got astride my motorcycle, and headed back to Chamcar-Loeu. I was in a rotten mood and wondered, as I was riding, if my staff had had any trouble during my short absence. I had no idea what kind of a mess I was going to find on my return. The Khmer Rouge and North Vietnamese assigned to the roadblocks had begun to know me by sight, so my trip was surprisingly trouble free. The news of my return quickly spread, and everybody on the plantation gave me a warm welcome. I passed all that day and part of the night talking, eating, and drinking with them. In fact, they had been afraid I would not return, and they shared their concern with me. For my part, I needed this warm reception to help me forget my deep sadness for at least a little while.

Plantation life kept on with its daily share of professional and extra-professional problems. A similar rotation, organized for a month later, ended badly. When landing at Chamcar-Andong, the wheels of the four-seater plane, flown by yet another pilot, struck the top of a levee that had been built across the landing strip to prevent the passage of ox wagons. Its presence shortened the runway and allowed only small planes to land. The four-seater overturned but, luckily, did not catch fire. We ran to help the pilot and his two passengers get out of the flattened frame. Miraculously, although two of them were injured, they were able to walk. The third was groggy but seemed apparently unhurt, which the plantation nurse later confirmed.

My two colleagues from Chamcar-Andong and I were very glad that the accident had not taken a tragic turn, but we were shattered because the plane was unusable and we were stuck in the communist zone instead of going to Phnom Penh for a month's leave. We were now six hostages instead of three. We at once negotiated for the departure of the three new arrivals by motorcycle first thing the next morning. The Khmer Rouge, agreeable to keeping their three familiar hostages, raised no objection.

I did not go back to Chamcar-Loeu the same evening because I had a heavy heart thinking about my little family, who were no doubt deeply disappointed when they had not seen the plane return. Instead, I stayed at Chamcar-Andong drinking all night with my compatriots. The next morning, when I got home to Chamcar-Loeu, with a raging headache, I took time to think about my future and that of my family. I drew several conclusions about the general situation at that time:

(1) The Vietnam War, in which the U.S. was then strongly involved, did not seem to be evolving quickly in the right direction.

(2) Cambodia's invasion by Khmer Rouge and North Vietnamese troops, in which the U.S. was not directly involved except for air attacks, no doubt had the potential to evolve toward a speedier conclusion than the situation in Vietnam. In only one month, half of Cambodia had already been successfully overtaken. The great paradox was that, for the North Vietnamese, an invaded Cambodia provided a safe and convenient base of operation for their troops; a communist Cambodia would have no need for the continued presence of those

troops. It was, therefore, to the advantage of the North Vietnamese to delay the Khmer Rouge advance.

(3) The situation on the plantation most likely would not improve rapidly. This conclusion was based on my first-hand knowledge of the very complicated situations on the ground all over these two war-torn countries, and took into account the increasing U.S.S.R involvement in the war (providing advisors, weapons, and materials).

(4) Finally, and most important, I was no longer a young, single man looking for adventure. I was now a forty-year-old man with a family.

All my experiences in Southeast Asia, my skills producing natural rubber, my knowledge of technical and psychological warfare, everything I had learned over fifteen hard years would be of no use in France or Europe. That left me with two career choices: staying in Cambodia, living in this mess, waiting for the war to end, not knowing how it would end, only knowing that I would have to leave anyway if the communists won; or going back to France as soon as possible and retraining for another profession. I chose the second way. Although I would have liked to share the decision making with Monique, I made it independently and confidently, knowing that she was sick of the hard life we had and that she wanted a peaceful, normal life for her family. From that moment, barring any unexpectedly favorable turn of events over the next few months, I planned to escape from the hell in which I was living and working.

Twelve

Escape

I had to keep my plans in the strictest secrecy until the last minute if I wanted to succeed. While I would have liked to have informed Monique, she was pregnant and didn't need any additional stress and worry. I, therefore, tried to continue to work and live as usual.

Over the next several months, the situation worsened intolerably. The Khmer Rouge and North Vietnamese increased their power and control over the people. Although their methods were different, their results were similar, dreadfully real, and sometimes just impossible to face. The more resistance the Khmer Rouge met, the more repression they used. Their methods were so much more harsh than those used by the North Vietnamese that the people soon began to prefer the foreign Vietnamese to their own compatriots, which incensed the jealous Khmer Rouge. In fact, the two communist allies had only two middle-range goals in common:

(1) the defeat of the non-communists, and

(2) the expulsion of the Americans.

The North Vietnamese long-term goal was, no doubt, to go back into North and South Vietnam in order to make it a unified communist state. The Khmer Rouge, who would not have been able to "liberate" Cambodia by themselves, tolerated the North

Vietnamese because they needed outside help. In addition, the concept of planning for the short-, middle-, or long-term haul was beyond the planning capabilities of the Khmer Rouge. They believed they had lost time by stopping the invasion. They became more and more impatient and angry. In all probability, they were already anticipating the waging of a genocidal war (so discreet that it would remain unknown to outside observers for some time). This unspeakable and brutish plan was put into effect just a few years later.

Thus, the Khmer Rouge increased their pressure with the blackmail, corruption attempts, kidnappings, arbitrary requisitions, and executions that became the routine of our daily life. They would come into my office and shoot hails of bullets into the walls, requiring impossible things such as several tons of rice, barrels of gasoline, extra food, involuntary "volunteers," and more. The situation was exhausting, particularly because I was having more and more trouble obtaining supplies. The fuel and the money, without which we could neither work nor live, were subjected to the horrific black market surcharges that nobody could control (but which benefited the Khmer Rouge either directly or indirectly). In order to isolate me more completely, the Khmer Rouge demanded that I dismiss some of my most trustworthy workers whom they disliked. As their demands were backed up by threats of violence, I had no option but to cooperate. They also wanted me to raise salaries, allowances, and subsidies for everyone in order to positively influence people who remained unsure about them. I suspected that the Khmer Rouge, who were usually more direct, had taken heed of advice from the North Vietnamese in the effectiveness of such psychological methods. These were not methods that they had used before. As these psychological methods took much longer than their usual methods of violence and coercion, the Khmer Rouge, unfortunately, lost patience and quickly returned to their usual bloody behavior.

The Khmer Rouge had some limited success in recruiting willy-nilly from among the population. Although most inhabitants were not ready to embrace the communist cause, some of these likable Cambodians, pressured by abominable means, were coerced into betraying others. For the Khmer Rouge, the results justified their brutal means. I knew several cases in which a man was recruited after the arrest of a relative accused of displaying a "non-revolutionary spirit." Such recruiting sometimes resulted in the release of the accused; other times the accused disappeared without explanation.

It was now out of the question for me to try to save everyone. Por Try's case had been decently defensible; these new Machiavellian cases were not. Leading my own investigations could have been counterproductive for the victims and even worse for me. In some cases, the Khmer Rouge forced people to denounce each other by using the children. By plying children, first, with unloaded weapons and, next, with ammunition, they easily gained control of the families. The North Vietnamese, on the other hand, contented themselves with occupying their zone, keeping a vigilant eye, fighting the planes, and controlling the people's movement.

The Malaysian ethnic group, about two thousand Muslims who lived together in one village on the plantation, tolerated this scary climate poorly. Some tried to escape and get back to their own country. Several succeeded; some were killed as an example designed to discourage others from trying.

Conveying the payroll from Phnom Penh worried me constantly as there were always one or two money sacks on the move somewhere in the country between the capital and us. Although I tried to continue to pay the personnel monthly, I requested my general director to send the payroll to me several times a month and by several different means. This solution decreased the risks of piracy, but it did not solve all the problems. The money was carried by at least two trustworthy faithmen with huge merit. The money bags were hidden in rice sacks, and then conveyed by ox cart, motorcycle, or foot. In order to avoid roadblocks and patrols, the carriers never followed the roads. Instead, they went through forests, rice fields, and rivers until they reached my office, which I kept open day and night to receive the payroll whenever it might arrive.

This worry, on top of all my other worries, became the stuff of nightmares, and I began to wear down from the daily torment I had experienced over the last several months of living in this bloody hell. Moreover, the pressures bearing down on my personnel had many repercussions and, as waves of dissent spread round, they seemed to intensify by the time they reached me. Even the subtle, firm, but derisive resistance I put up against the Khmer Rouge began to get the better of me.

Despite my best efforts, rubber production slumped. I felt as if I were constantly trying to fill a leaky bucket. I could not see a safe way out of this nasty quagmire. At last, my loneliness (unbearable since the birth of my son Jean-Luc in Phnom Penh three months before), the heavy stress that weighed on me, the blocked future, and my fear of a possibly tragic ending to this drama, all combined. I thought more and more about how I could escape.

My decision to leave was very critical for my company as well as for myself, so I decided not to do anything without first informing my boss. I accomplished this by means of an exchange of coded correspondence with Phnom Penh. I was relieved when I received word of the general director's understanding of and agreement on this matter. In fact, I learned a little later that the company was already very satisfied that my two colleagues at Chamcar-Andong and I had managed to stay as long as we had in the communist zone while having to endure such hellish living conditions.

The great mysteries were how to avoid throwing the two plantations and their populations into turmoil, and how to facilitate the continuation of work until we would be able to come back, whenever that might be. We had to go on behaving as usual but, because of the great mutual confidence I had with my two Cambodian assistants after our several years of working together, I took the risk of informing them. They would be in charge of running the plantation and producing rubber after my departure.

No doubt, their hardest task would be explaining to the Khmer Rouge and, to a lesser degree, the personnel and their families why I had suddenly left. We agreed to say that I was lonely from being stranded so far from any other Europeans and anxious to see the son I had never seen. The Cambodians would understand these reasons perfectly; the reaction of the Khmer Rouge was less certain—they were not known for their understanding, humanitarian nature.

Maintaining the strictest secrecy, my two colleagues from Chamcar-Andong and I focused on working out the details of a feasible plan of action. Traditionally, our company had a policy of favoring and helping family plantations in the neighborhood, and there were many family plantations near us whose latex was treated at our plants. From time to time, we visited these small plantations in order to give the owners technical advice. It seemed appropriate, on our personal D-Day, for the three of us to undertake one such visit/consultation to the family plantations located to the south and southwest of our plantations. That would place us reasonably close to the "border" that separated the communist zone from the rest of the country. In order to allay any suspicions, we both officially and informally announced this inspection tour to the invasion troop leaders and all over our two plantations.

I passed the most important part of these last crucial days with my two assistants, Nou Try and Por Try (who, in spite of their common name, were not brothers). We supervised the different loggings together, and I did everything humanly possible to give them all the details I could about each field. The evening before my escape, Nou Try, and especially Por Try (whom I had saved from the clutches of the Khmer Rouge), and I could not help but weep our eyes out. These two men where going to feel like orphans, and we all tried to put our minds at ease by talking about anything and everything. I could make no distinction between liking and loving; these men were brothers to me.

Of course, I could not sleep all night. I made it a point of honor to be present at the 4:00 A.M. roll call and go through all the bleeding shares by bicycle. My heart was broken when my workers, who knew nothing, said "Hi" or made a friendly gesture; I was deeply touched by these signs of friendship at this crucial moment of my life. Nevertheless, I could not help but feel like a devious character, a false friend who put up a pretense in order to save his own skin and, although I felt sick at myself, I was overflowing with joy at the thought of seeing my family again if all went well. My God, how difficult this life was!

Anyway, our escape plan was ready and our futures at risk, to say the least. At about 8:00 A.M., my two French colleagues arrived at my office, both of them riding the same motorcycle and seeming just as strained as I was. After reviewing all the possible scenarios with which we could be faced and checking our Khmer Rouge and North Vietnamese passes, we started to the first family plantation.

These plantations were not located along the road but were scattered everywhere within a twenty to thirty mile radius of Chamcar-Loeu, whereas the demarcation line was about forty miles from us. As long as we stayed inside the perimeter of the farthest plantations, we could justify our presence; once outside that perimeter, we could not. It was then that we would face the greatest risk. Patrols from the two military camps kept a close watch over the area, and we knew from my smugglers that ambushes were often set. Our biggest problem was that we did not know where exactly this fictitious "line" was located, and my guess was that the troop patrols felt the same confusion. Thus, we could meet anybody, anywhere, at any time outside of our perimeter and the situation could be dangerously ambiguous, especially if the soldiers were either nervous or trigger-happy.

Farther along, we decided to split up, my two colleagues remaining together and me going on alone. We hoped that it would reduce suspicions if we should be arrested, and increase our chances of succeeding. We did not know at that time whether it was the right or wrong decision, but it represented our best guess of what would work. After exchange wishes of "good luck," we started off in our own directions.

I actually found my visit to the first two or three plantations very interesting because the peasant owners, who had met me many times at my plantation when they brought their latex, had never or almost never entertained me on their own plantations. My presence on their plantations was a novel convenience, and they showered me with technical shoptalk to the point that I almost forgot the real purpose of my being there: escape. While on these visits, I met only one Khmer Rouge patrol, which examined my pass and left me alone.

Time flew quickly as I made my way to the border. It was about 5:00 P.M. when I finally was about ten miles away from my goal. The farther I was from my own area, the less I knew about the terrain. I didn't know what kind of obstacles I was going to encounter, and I wondered if I should keep my motorcycle with me or not. I didn't need it to cross the approximately ten miles left to reach the dreaded line, and its noise and bulk might prove to be a handicap. On the other hand, after crossing out of communist territory, I would still be about eighty miles from Phnom Penh. For that stretch, the motorcycle would prove a very handy thing to have indeed. Smugglers had informed me that the Phnom Penh-Kompong Cham Road was not frequented since the invasion, and it would be safe to use the motorcycle there. After a minute of thinking, I adopted a compromise solution; I decided to push my motorcycle. It was very heavy (350cc), but I felt more secure. If I met a communist patrol, I would crank it up and dodge bullets as I fled.

I had barely reached the halfway point before I began to feel exhausted from lack of sleep and the exertion of pushing the motorcycle. I was following a trail through forest and savanna, following the sun, and I pushed on, with jaws clenched. Suddenly, I heard the purr of a reconnaissance aircraft so close to me that I barely had time to throw my motorcycle and myself into a copse. The last thing I wanted was for the pilot to see me, especially if there were Khmer Rouge or North Vietnamese patrols in the area. I waited, scared stiff, while the plane passed over the trees, directly above me. Nothing happened, and I was able to progress again, with difficulty, for another hour. It was now about 7:00 P.M.; darkness fell at about 8:00 P.M. at that time of year. I began to be worried about my adventure. I had no idea exactly where I was—whether I was still in the communist zone or not. I was not sure if I should advance straight ahead through the forest or turn southeast in order to get back to the road.

I stopped for a short time, well camouflaged, in order to rest and think it through. Almost immediately, bullets were flying over my head. I was frightened out of my wits to discover that I was between two enemy patrols whose mutual fire was crisscrossing above me! This was the biggest scare of my life. When I had faced the Viet Minh, I at least had had a weapon. Here I was a helpless, vulnerable, and unwilling "participant" in a firefight. It was unbelievable; I was enraged. So much for the luxury of a little rest!!

Nevertheless, my luck held even in the midst of this immediate peril. I now knew where I was; I was lying almost on top of the line. However, I was far from safety unless the two patrols stopped firing away at each other, and did so without finding me. I begged God to grant my prayer and, the more I prayed, the more I tried to become smaller and smaller. Furthermore, I fervently hoped that none of the opposing soldiers would have the bright idea of charging the other side; they would have charged across my abdomen!

The racket of the shooting drowned out the sound of two bullets that ricocheted off my motorcycle, and I miraculously survived the shooting. However, my ordeal was far from over. The soldiers had stopped shooting, but I didn't know if they had gone away or had hidden in ambush. After waiting through a long period of silence, and with pins and needles everywhere in my body, I tried to convince myself that staying put in the deepening darkness was stupid if the two opposing patrols had already left. I decided to try a trick I had seen years ago in a movie. I crawled around and found some big stones to throw in the direction of the two patrols, figuring that if nobody shot I probably was alone in the middle of the savanna, waiting like an idiot. I threw my stones.

Nothing happened. I got up slowly. Still nothing happened! I was beside myself with joy despite the reality that I still had to reach the road to the south, pushing my motorcycle and risking interception by a Khmer Rouge patrol along the way.

After another long hour of difficult progress, luck favored me again. I came across a Nationalist post controlling this section of the border. I arrived at a most awkward moment. The soldiers who had shot over my head were telling the others about their battle and crowing over their victory. They were very surprised to see me there and, although I was completely worn out, I had to tell them about my odyssey while drinking a warm beer that they kindly offered me. I tactfully omitted the episode about the shooting and left them their vaunted "victory"—which I knew was neither a victory nor a defeat, but a mere skirmish undertaken, thankfully, from a distance. Anyway, I was happy and tired and they were happy and kind, so I remained silent about the unimportant stuff. I only wanted to ride my motorcycle on to Phnom Penh, clasp my family in my arms, and meet my son for the first time.

The road was deserted and I beat the speed record for covering the distance from the control point to Phnom Penh. Along the way, obstacles continued to thwart me. The Mekong ferry was closed for the night. Undaunted, I woke a fisherman and paid him a grand fee to take me across the river, but I lost two more precious hours. I reached Phnom Penh in the middle of the night and the immense happiness that overcame me as I fell into my family's arms was indescribable and immeasurable. The next few days seemed like being in heaven after having endured hell. I rejoiced to learn that my two friends from Chamcar-Andong had reached Phnom Penh via another route about an hour before I did. They had made their way without incident and, not having run into a Nationalist post, had gained time on me while I was recounting my adventures to the celebrating soldiers. Our nightmare had ended happily. All three of us were now safe.

Even though everyone was pessimistic about a return to the plantation, the company asked me to stay in Phnom Penh for two more months, hopeful that the situation would change in such a way that I would be able to go back. According to the information we got through smugglers and the bamboo phone, communist troops had advanced a little bit on every front in Cambodia, and the Khmer Rouge had begun to commit bloody repressions in several places. The Khmer Rouge could not have waged this war without their allies, the North Vietnamese. The U.S. Air Force and the Cambodian Nationalist troops would have been too strong for them to handle alone. While the North Vietnamese were waiting for a fissure to develop in the American-South Vietnamese wall of resistance, the Khmer Rouge, crazed by their enforced wait, found nothing better to do than execute people on fallacious pretexts.

After two months, my family and I left Cambodia with heavy hearts. We were officially taking a three-month vacation in France but, given the conditions at the time we left, I knew that I would never see Cambodia again. That knowledge filled me with a deep sadness.

Even though I had not yet received my letter of dismissal, I began to look for a job two weeks after our arrival in France. The chance of returning to any location in Southeast Asia was very slim. I wrote hundreds of letters to companies all over the country and Europe, hoping to work for a French or British Company, perhaps in Africa. My experience and my ability to speak two Asiatic languages, although original, impressed nobody. I received my official letter of dismissal at the end of a fruitless four-month job search. Just as I began to worry, I chanced to meet a manager who trusted me after he heard my stories about Vietnam and Cambodia. Although I did not have an appropriate diploma in the field, I was hired to work as the personnel manager of a French National company involved in the production of chemicals and explosives that employed about eight thousand people. Over the next three years, I made it a point of honor to study engineering organization and the psychology of communication, fields in which I received diplomas, in order to increase my competence in my new profession.

Meanwhile, my family got used to living in France. Surprisingly, they adapted better and more quickly than I did, although all of them had been born in Asia. During these three years, my wife and I followed the evolution of the Vietnam and Cambodian wars through the media. We experienced, at a distance, the fall of Phnom Penh and Saigon on the 30th of April, 1975. We were deeply affected by these events, knowing that our American friends would never have lost this war if their Vietnamese non-communist allies had been motivated and honest with them.

My love story with Southeast Asia had gone through many ups and downs during the two decades I spent there. It had started with war, moved through a wonderful honeymoon period that brought success in work, friendship, and love, only—sadly—to end in another war. In spite of all I experienced, I have to say that I feel drawn toward Vietnam and Cambodia as much now as when I was young. It is a romance that has never ended, and Monique and my children are with me to forever remind me of those days. Perhaps my soul will one day fly over these beloved countries where I left such a large part of myself.

Epilogue

It is now the year 2000 and I am seventy years old. I have been living in the United States for eight years striving to learn English, something that I was not able to do when a young boy in school because the Nazis, who occupied France when I was ten, forced us to learn only German as a second language. I am also helping my youngest son, Jean-Luc—born in Phnom Penh while I was trapped in the communist zone and now thirty years old—to finish his graduate studies. A few years earlier, my daughter, Dominique, who had languished in Phnom Penh with her mother while I was in danger, spent six years in the United States in order to complete a Master's degree in English, which she now teaches in France.

In addition to working on my English, I have become a member of the United States' Veterans Association of "In-Country Vietnam Veterans," and have been proud to stand alongside my fellow veterans in various ceremonies and parades over the years.

I have dedicated this book to those Americans who gave their lives in Vietnam, and also to all those who survived the war but who were wounded in body, mind, or spirit. In my opinion, the Americans did not fail in Vietnam. They championed a just cause, but they were linked to a powerless political ally and, again in my opinion, politicians and military strategists never get along well with each other, whatever their nationality.

Only the intermingling of experience with reflection can impart wisdom. I believe this to be true for everyone everywhere. My hope is that the wisdom distilled from the two consecutive wars in Vietnam, the French and the American, and from the war in Cambodia, will bear positive fruit in the future.

Appendices

Appendix One

How Rubber Is Cultivated[1]

A working rubber plantation in either Vietnam or Cambodia would be located where the landscape is already very densely forested. Both these nations are located in a topical environment that has a very strong monsoon season. It is normal to see two meters or more of water during the rainy season, which lasts about six months each year. The rain striking the parched tropical ground in June heralds the end of the dry season. Heavy rains and violent thunderstorms continue until November. The daily deluge of water quickly reaches all parts of the plantation and its villages; no storm drains are sufficiently able to drain it off. Just as suddenly, usually about an hour after the storm, the sun is shining and the water has been totally absorbed into the ground.

The heat is still strong; the humidity unbearable. No one can raise a finger without sweating. Air conditioning is useless because everyone works outside even in the midday heat. Rivers are overrun with fish; rice fields and ponds with bullfrogs. Women no longer need to stand in line at the well because rain water is systematically drained into buckets by gutters (made of bamboo

[1]The information in this section reflects the management system and working conditions that were the norm during my years as a plantation manager in both Vietnam and Cambodia. The way in which one cultivates and harvests rubber has changed little. However, with the nationalization of these plantations, both the style of management and the working conditions on the plantations themselves have changed considerably.

split lengthwise) around the roof of each house. All available containers are used to gather and store this precious water. The daily diet of the dry season (black pig, dried fish, and rice) now changes to fresh fish and fish products, fresh vegetables, and rice. In general, life becomes a lot easier for the hard working people of this region.

Vehicles assigned to transport rubber toil along muddy tracks that cover hundreds of miles and run alongside levees that connect the various areas that make up a plantation. This is also the season when tropical diseases emerge, especially malaria. Nobody sleeps without a mosquito net. In the jungle, dysentery menaces those who fail to purify their water before drinking. European pioneers sometimes put cognac or scotch into their water before drinking it, which is why others derided them as the "saturated." Other common diseases are more on the rampage than usual.

All the vegetable species, wild or domestic, proliferate with an abundance never seen in Europe. Moreover, there are many impressive trees in the jungle. Those called "cheesers" have a circumference of up to thirty-five feet. Between the trunks of these large trees, which seem to point endlessly up to the sky, there is a tangle of smaller trees—a difficult obstacle to penetrate.

Before deciding on a location for a plantation, a technical team comes into the area to take soil samples and make an environmental analysis. The culture of hevea rubber trees requires a lot of water for two reasons: (1) the roots of the hevea need to draw from the water of the phreatic (groundwater) strata during the dry season, and (2) the manufacturing process that transforms latex into rubber requires a lot of water. In addition, when rubber trees are in yield, the trucks transporting the latex must be able to move with a minimum of difficulty over the levee tracks (which are laid out every 275 yards in a north/south–east/west grid pattern all over the plantation). Therefore, rubber plantations are always located in a flood plain close to a river.

Clearing the Plantation Area

The first task is to clear the ground. Day and night, during the dry season, trees are cut down and then pulled out by powerful bulldozers (the ground is not overly saturated with water from December to May, and the bulldozers can work to their full capacity). Tree trunks are cut and stacked in huge lines. After a few months of drying in the sun, they are burned. As soon as the fires are lit, dry leaves and small and medium branches burn like paper. However, the huge trunks can take several months to burn and, if the process hasn't finished before the rainy season sets in, the rain extinguishes the flames. At times, some cut trees are given to the villagers, who use them to build houses, sheds, or furniture.

The cleared ground is divided by the levee tracks into sixty-acre plots, on average. The plots are listed and numbered. The levee tracks are useful for traffic; at the same time, they protect against flooding and erosion.

The plantation evolves, grows, and extends each year as long as there are appropriate areas available to clear and plant. After forty to fifty years, when the oldest planted areas become nonproductive, the old hevea are cut down and the areas are replanted with young heveas or hevea seeds, thus restarting the cultivation process.

Plantation Villages

A high performance plantation must be autonomous. Electric power, human resources, health, religion, and schools—as well as all the appropriate equipment, buildings, and other installations—are integral parts of the plantation. All employees, from executives to laborers, live on the plantation with their families: wife, husband, and children, of course, but also parents, brothers, sisters, and sometimes nephews, nieces, and grandparents. This has the disadvantage of increasing the accommodated population, but the advantage of creating a strongly stable environment. The resultant decrease in job turnover means that the work proceeds in a more orderly fashion.

A plantation can have from three thousand to ten thousand people living in one to three or more villages scattered over the plantation. The primary one is the center or central village. Located as near as possible to the river, it contains the factory, power station, mechanics' workshop, automobile and truck fleet, warehouse, offices, and houses of some of the workers. In the other villages, employees and their families are divided into living groups by religion (e.g., Catholic, Buddhist, or Muslim), and each village is equipped with its own school and an appropriate place of worship.

It takes six to seven years before the seed that has been planted can produce the first rubber harvest. The completion of all the installations must be coordinated so that rubber production begins smoothly. For example, in the clearing phase, the planter must plan and supervise the levee track construction—a total of more than 3,000 miles for a 13,000-acre plantation. He must take into account the impossibility of constructing levee tracks during the rainy season. Also, it is difficult to construct buildings and installations during the rainy season. That makes the dry season a period of intense activity.

Although the plantations are often isolated, hiring workers is not difficult. New workers are often located by word of mouth, with current employees recruiting others from their home areas. The jobs are attractive because they pay a reasonable wage. In addition, workers receive free housing, two pounds of free rice per workday, and other allowances in relation to the size of the family. The plantation also provides free medical treatment to everyone, and the company doctor comes once a month to examine the more difficult cases. The company's plane can also evacuate emergency cases.

Plantation Management

Each plantation is managed by a European director who has two, three, or more assistants (dependent on the size of the plantation), most of whom are also European. In general, there is one assistant per 2,500 acres of yielding cultures, plus one mechanic assistant. The most experienced of the assistants functions as assistant manager, and will replace the director if necessary. The mechanic assistant lives in the center village. He is responsible for the management of the vehicle fleet, the power stations, the rubber factory, and the warehouses (which contain everything from spices and rice to all the assorted equipment essential to the production of rubber).

The administrative office is run by natives supervised by the director. This office is responsible for payroll, general attendance, determination of rice rights, material orders, and receiving. The most important problem, especially during troubled periods, is maintaining the chemical supply from abroad. Rubber cannot be manufactured from latex without formic and acetic acids, and ammonia.

Instead of a telephone, each plantation communicates with the outside world by radio. Additionally, each plantation maintains a grass airfield so that a small company plane can bring in medical supplies and the doctor, and airlift the sick and injured.

Hevea Culture

In hevea culture, there are two parts that are very different from each other: (1) the young cultures, which are also called extensions, and (2) the yielding cultures. Seedling trees, or cultures, yield latex after six years; grafted trees after seven. If the clones are well selected, grafted trees produce much more latex than ones grown from seedlings.

With the exception of a few special cases, an extension project rarely consists of less than 300 acres because of the considerable commitment required in equipment, material, and personnel. The more acreage developed, the less the cost per acre.

Depending on the weather, the condition of the soil, and the rate at which the land is cleared, the construction of the levee tracks can begin three to six months after the land clearing process has started. The levee tracks are generally seven yards in width, and five to six feet in height, packed down. Preferably they are wide enough to allow two-way traffic, especially when the cultures are yielding latex. When the planned planting zone is uneven, both the levee tracks and the hevea planting are laid out in accordance with level curves or terraces in order to guard against erosion during the rainy season.

Once the clearing, soil preparation, and levee tracks or dikes are finished, usually May or June, a general sowing of couch grass is made. This will protect the ground from the sun during the dry season and retain moisture. Only the top of the levees, where traffic is heavy, remain unplanted. The area will then be kept fallow for eight to ten months, until the arrival of the next rainy season.

The yielding cultures are seeded in December and January. One must create some seed nurseries and, in order to give the young hevea the best base possible, select the seed from clones known for the robustness of their taproot. The hevea seed is about the size of a walnut, and it has a dark brown color and very smooth shell. Several hundred thousands of seeds are collected and selected by women and teenagers working in carefully trained teams. Indeed, this task is crucial to the degree of success one will obtain from this planting.

After having been selected, the seeds are soaked in an ant pesticide (to combat the number one enemy of newly planted seeds) and arranged in germinating huts. It takes two to four weeks for the seeds to germinate, and a bonus is given to the personnel who sorted the seed based on the percentage of germination.

During the delicate and important soil preparation phase, the area to be planted must be marked with stakes that indicate the future site of the heveas. These stakes must be laid out in perfect lines—north to south, east to west; horizontally, vertically, and diagonally—in order to maintain a strict squareness within the area. The distance between each stake is determined by the abundance of foliage on the clones, the resistance of the tap root, and the average height of adult trees. Common dimensions are 3.5 x 8 yards, 4 x 6.5 yards, and 5.5 square yards.

When the stake marking is finished, a strip of ground two yards wide must be perfectly cleaned all along the north-south stake lines. This work must closely precede the planting, preferably being done the day before, so that the seedlings will break through the soil before the couch grass resprouts and suffocates them. As the seeds from the nursery have already started to sprout, they are carefully carried to the site in flat-bottom baskets and planted, following the north-south lines, at the rate of seven or eight seeds per staked-out site. They then are covered by hand with about one inch of soil.

During the planting season (April or May), everyone hopes for the timely arrival of the first rain that will guarantee the water necessary for the success of the planting. Planting is done as close to the end of the dry season as possible, but the rhythm of the rainy season is not predictable. Sometimes the season is delayed for months or, more rarely, for a full year. If the rainy season is delayed, one is forced to water with tank trucks that haul water to the site from the center village, a distance of several miles. This is a long and arduous task, done at night to avoid temperature extremes and excessive water loss through evaporation. Rain at the necessary time ensures higher yields and, therefore, higher profits at harvest.

If all goes as planned, the young plants grow out of the soil some two or three weeks later (in June or July). Four or five out of the seven or eight seeds planted at each site should have thrived. If not, replacements must be made. Line cleaning must be done every forty days during the rainy season and less often in the dry season. This work is done by piece-work teams who are paid in relation to the number of young plants cleaned vis-à-vis the number broken.

Once the young plants have developed to about one inch in diameter at the base, it is possible to proceed to the grafting stage (usually during October and December). This difficult work is done by specialist teams that move from one plantation to another, all over the country, during the grafting season. These teams are remunerated according to the percentage of graft success. Supervising the process and checking the results for the twenty to sixty days after grafting becomes a constant and demanding task during the three-month grafting season. Both these stages are very important in the hevea culture process because the planting success rate, and then the grafting success rate, determine future production.

About two months after the grafting, when the graft eye forms a new growth, one must be very careful that each tree grows without branches to about ten feet high. This will determine the quality of the future bleeding panel. During this time, one also must continuously guard the entire area, especially at night, in order to keep animals, particularly elephants, away from the fragile and sensitive trees.

The goals of the grafting process are to get hevea with smooth trunks that favor a good bleeding and strong crowns at a good height that resist the wind. One must avoid the need for replacement trees because any holes in the line make the trees vulnerable to wind and thunderstorms. During the next six to seven years, cogent care, treatment, and attention are lavished on these newly grafted extensions. A serious, thoughtful job is the best guarantee of a healthy and profitable crop.

Bleeding Preparation in Young and Virgin Cultures

The preparation for bleeding is made during the one-month bleeding stop in the yielding cultures (this stop is necessary both because of the waiting period needed between the preparation and bleeding of the many bleeders available and because of natural defoliation). A vertical line, positioned to the southeast (away from the usually prevailing direction of the rain), is drawn from ground to crown on each tree, using a six foot wooden ruler and a metal hook for scoring the tree. From this line a full-sized metal spiral is applied, and the site of the future bleeding notch is marked. For the full spiral to be well applied, about one and a half feet must separate the top from the bottom of the notch so that it neither slopes too much nor is too flat. The bottom of the notch must be two and a half feet from the ground, and the gutter must be located four inches lower than the bottom of the notch. After the drawers have completed this step, the dry bleeders move through the trees and mark the notch definitively with bleeding knives, straightening it if needed. This allows the bleeder, when he bleeds for the first time, to find a good slope with a minimal loss of time. When the trees have been counted, listed, and shared out in alternates, the bleeding can start, and the new extension can be put into full production.

Bleeding the Yielding Cultures

In order to get the best production possible and prevent either the deterioration or death of the vegetable material, the rubber tree, hevea, must not be bled every day. A rhythm adapted to the age of the tree must be chosen. With a full spiral cut, one bleeding every three days (D3) is the usual rhythm for seven-to-thirty-year-old grafted trees. A two day (D2) reversed semi-spiral bleeding with higher notches can be used for some trees, more than thirty years old, when the bark near the base of the tree has become unbleedable. A bleeding process of full spiral or semi-spiral cuts on opposite sides each four days (D4) can be used for grafted young trees, seven to twelve years old. However, young trees are particularly fragile and sensitive to bark panel diseases.

In relation to bleeding rhythm D3, more common in my company, the yielding cultures are shared out in three parts that are called alternates A, B, and C. A is bled the first day, B the second day, and C the third day, rotating all through the year, without any stop except when the leaves are falling during the dry season (February). All the trees—which for a 12.5 thousand acre plantation equals about 2.3 million trees, depending on planting density—are counted and listed in a special log. Each alternate is divided into shares, with one share consisting of three hundred to five hundred rubber trees, according to the age and size of the trees. Each worker, called a bleeder, always bleeds the same share in each alternate. The bleeders are divided into twelve to fifteen teams, each with their own team leader; team leaders, organized in groups of ten, are themselves responsible to a supervisor.

The technique of bleeding requires subtle manipulation and skill. The latex is harvested by cutting a thin strip all along the notch marked below the full spiral that runs around the tree. The best results are obtained by bleeding the thinnest possible strip each time in order to ensure harvesting the most latex while consuming the least bark. For example, a good one-year bleeding consumes about 3.9 to 4.7 inches of bark for a skillful bleeder and about 5.9 to 7.8 inches for an incompetent one. With the skillful workman, a bark panel of about three feet lasts five to seven years. When the capital disappears faster on one side, the bark doesn't have enough time to rebuild itself when the notch is raised on the other. Regular bleeding is an important criterion of health for a rubber tree, so all of the workers on a plantation, manager included, rest only when the trees rest during the bleeding season.

During the annual lull from bleeding, the trees are counted and divided up again because some trees have died from sickness, fire, or other catastrophe during the preceding year. During this period, huge quantities of dried leaves are swept and burned in order to avoid an uncontrolled fire. In addition, the young hevea planted seven years before are prepared for the next bleeding.

Moreover, sick trees are treated. The organization of the bleeding season described above allows for the efficient crop control necessary for successful results (such as maximum production; minimal bark consumption; limited tree damage; maximum cleanliness of notch, gutter, cup, and ground; and well-shaped notch).

Bleeding must be done early in the morning because the latex bark is very susceptible to coolness. This holds true for all seasons except the rainy season when, if it has rained the night before, one must wait for the trees to dry (if they are wet, the latex might flow onto the ground instead of into the cup). The latex flows all along the full spiral notch into the small gutter that has been hammered into the bark above the cup, which will hold about one quart of latex. This cup is suspended from a metal support tied around the tree.

A monthly rotation of bleeding control for each tree is maintained. First thing each morning, using a small borer set to bleeding depth and some colored chalk, an assistant marks bark consumption and notch slope. The results of this monthly control for each bleeder is written down in a specific booklet and the grades are increased or reduced in relation to the quantity of gathered latex. A monthly bonus is paid to each worker in addition to their salary, and a good bleeder can receive a bonus equal to as much as one-third of his salary.

At about 11:00 A.M., the latex is gathered by each bleeder by means of a rubber cleaner and some metal buckets. It is important to replace the cup on the metal support correctly after the gathering because the latex continues to flow slowly for a few hours. In the rainy season, the latex gathering is often a race between the bleeders and the rain. To limit losses, some workers are assigned to watch the clouds and the direction of the wind from strategic points around the plantation. These warn of the coming deluge by beating on large, hollow bamboo trunks. The bleeders, hearing the signal, rush to gather in their buckets before the rain starts. Sometimes it rains before the end of the bleeding, so everyone must wait for the trees to dry and then bleed them again. Other times the bleeding is partly or completely lost. In these cases, one tries to make up for the loss over the next few days.

When the gathering of latex from the cups is finished, the bleeders carry their buckets, team by team, to a special collection point where the team leader checks and records the volume brought in by each bleeder. At about noon, the plantation's fleet of tank trucks collects the latex and transports it to the factory, usually located in the central village.

Of course, even after collection has finished for the day, latex continues to flow into the cups, more or less in relation to the temperature. If the weather is warm (but not too hot) and dry (with no threat of rain), latex can be collected twice a day. Normally, however, the flowing latex is allowed to coagulate, and the cup, notch, and ground coagulum are gathered just before the next bleeding, which usually takes place three days later.

When two collections per day are unnecessary, the afternoons are used to weed, to treat the trees for sickness, to treat bark panels after bleeding injuries, and to eliminate any dead trees that have fallen (and which hamper the bleeding of the surrounding trees). There is always a fresh outbreak of sickness among the trees during the rainy season, and the youngest workers become acrobats with ladders as they apply pesticide to branches hit by disease. Finally, with the approach of defoliation, dead leaves are swept up in piles that will be burned at night when it is cooler. This work is an assistant's nightmare. Fire is the worst enemy of the hevea and, if the dead leaves burn to within three feet of a tree, it is gravely damaged: the latex flows, the bark swells and warps, and the bleeding becomes impossible to check. Considering the hundreds of thousands, even millions, of trees and the surfaces on which this sweeping must be perfectly done, one can begin to comprehend the magnitude of this nightmare.

The Manufacture of Rubber

The transformation of latex into rubber is completed at the plantation factory, always located close to water and as near as possible to the yielding cultures in order to avoid lost time, which is always baneful to production quality.

When the team leader checks quality at the gathering point, he adds a water-ammonia solution (NH_5O) to the latex to prevent premature coagulation—a mechanic assistant's nightmare, especially in the rainy season. The tank truck collecting latex must travel between the different collection points and the factory as fast as possible for the same reason. It is not rare to see a tank truck skate along the levee tracks or get stuck in the mud on the bottom of a levee during the rainy season—a driver's nightmare. If the liquid latex stays in the tanks for more than a few hours, it cannot be transformed into first quality rubber. It will have to be labeled third quality rubber, with a comparable loss of profit—the company's nightmare.

At the factory, the latex is poured into huge, ten- to fifteen-thousand-gallon vats, where it is mixed with water (the necessary amount having been determined by a dry rubber content analysis) and either acetic acid ($C_2H_4O_2$) or formic acid (CH_2O_2). The mixture must be perfect in order to prevent such disasters as precoagulation or no coagulation at all. This mixture is quickly poured into smaller, five-hundred-gallon vats. Wet aluminum plates—wet in order to prevent the mixture from adhesing to the plates—are plunged into these vats, and the contents are left to settle.

It takes ten hours (more or less) for the contents to coagulate, at which point the aluminum plates can be removed. The placement of these plates in their frame is such that, once removed, the coagulum in the vat can be unrolled in one continuous strip (2 inches thick by 20 inches wide by 200 yards long). To prevent the coagulum from becoming dirty, the strip is unrolled in water. It then passes through a system of rollers where it is compressed until it is only some 2 to 3 millimeters thick.

Moving further along the conveyor belt, the strip passes under a large blade where it is sliced into thin sheets of rubber. The sheets of rubber are placed on drying trolleys, which are then locked in large buildings and wood smoked for three or four days until completely dry. Once dry, each sheet is hand inspected, sorted by grade, and packaged for sale overseas.

Rubber Classification From Top Grade to Lower Grades

(1)	RSS 1	Light, dried sheets of rubber with no black dots, bubbles, bamboo traces, mold, wet parts, or still sticking parts.
(2)	RSS 2	The same as RSS 1, but sheets with a few bamboo traces, very light rust traces, very rare small air bubbles, and/or a few black dots can be included.
(3)	RSS 3	The same as RSS 2, but sheets with air bubbles and some traces of bamboo can be included.
(4)	RSS 4	The same as RSS 3, but sheets with many large air bubbles can be included.
(5)	Cutting A	Perfectly dry, small pieces (as opposed to sheets) of RSS 1 and RSS 2.
(6)	Cutting B	Perfectly dry (perhaps even smoked too much) small pieces of RSS 3 and RSS 4.
(7)	Crepe	Rubber with a wrinkled surfaced that is produced from the cup, notch, and ground coagulum salvaged three days after each bleeding.

Rubber Tree Extensions

Seedling trees or young cultures, also called extensions, will yield latex after six years; grafted trees will yield latex after seven. These extensions are about three years old. Note the use of grasses and leaves to protect the ground from the sun during the dry season and to retain moisture.

Grafted Rubber Tree
[LEFT]

This is a nine-year-old tree that has been bled for two years. Note the marks of the full spiral cut. A notch at the bottom of the spiral carries the latex down the tree to the gutter which, in turn, deposits the latex into the waiting cup.

Seedling Rubber Tree
[RIGHT]

This is a fourteen-year-old tree that has been bled for eight years. Note the spiral, notch, gutter, and cup.

A Rubber Plantation Factory

The foreground shows five-hundred-gallon vats of latex that has been mixed with water and either acetic or formic acid. On the left in the background are the drying trolleys loaded with thin sheets of rubber. In the center of the background are the buildings in which the latex sheets are wood smoked for three or four days until completely dry. Once dry, each sheet is hand inspected, sorted by grade, and packaged for sale overseas.

Worker's Houses

Foreman's House

Appendix Two

Geography of Vietnam and Cambodia

Vietnam

Vietnam, as a reunited nation, is bordered by the China Sea along its southern and eastern borders with a coastline of about 1600 miles. China borders it to the north and Laos and Cambodia to the west. Its surface area is about 80 million acres, and its population in 1992 was 67 million. The tropical climate, wet and hot, works on fertile soils that produce dynamic vegetation: the dense forest.

Geographically, Vietnam resembles a picture of a "shoulder scourge" (the individual and ancestral means of carrying anything) whose two extremities, north and south, are the Vietnamese "rice sacks": the Song Hong (Red River) Delta on the north and the Mekong Delta on the south.

Mountains cover two-thirds of the country: the Annamese Cordillera, or Annam Mountains, look to the east and dominate the China Sea. The highest point of this range is Chu Yang Sin at about 8,000 feet. The Hauts Plateaux (High Plateau), the southern part of the cordillera, spreads across about 160 miles of the north end of the Mekong Delta. The Cochinchine Embankment, from Tay Ninh (the center of the Caodaist religion) to Vung Tau (Cap St. Jacques), is located in the extension of the Cambodian northern plain. This embankment was very hazardous in the past because of the presence of the fearsome *Anophele balabalensis*, a dangerous mosquito that entered homes but never landed, and so was almost invulnerable. Fortunately, since 1921, the founding of many rubber plantations has decreased the population of this deadly pest.

In 1964, there were about 300,000 acres planted in rubber trees in South Vietnam. Of these, two-thirds were yielding cultures that produced 75,000 tons of latex per year. The rice and hevea cultures are the most important exportable crops in Vietnam, but there are also tea and coffee plantations on the Hauts Plateau. In 1950 the surface of the rice fields was about 6 million acres, but the area yield was poor because of the war. Nowadays, the area yield reaches more than a ton per acre.

Cambodia

Cambodia is bordered by the China Sea along 320 miles of its southwest border. It is bordered by Vietnam on the east and south, and by Laos and Thailand on the north. Its climate is the same as Vietnam's, and its surface area is 50 million acres. The first census in 36 years was undertaken on 3 March 1998 by the Cambodian Ministry of Planning's National Institute of Statistics with financial and technical assistance provided by the United Nations Population Fund (UNFPA). The preliminary data estimate the population as falling just short of 11.5 million.

The name "Cambodia" comes from the Sanskrit "Kambuia," and is, so the legend says, a gift from the 2,500-mile-long Mekong River. Its plains exist only because of the river's cycle of flood and abatement. Each October, when the Mekong crests, it covers all the rice fields along its path. In Cambodia, the waterways that combine to form the Four Arms Plain swell to create a small, and generally stormy, sea. Tonle Sap Lake, which is usually fed by the Mekong River, reverses its flow, flowing into other lakes located to the north and covering about 8 percent of the country with water at depths of up to 35 feet.

Geographically, Cambodia is a large depressed zone surrounded by a semicircle of mountain ranges that form a natural boundary between Cambodia and Thailand: the Elephant Range in the southwest; the Cardamom Mountains (named after the spice) and Phnom Aural, its highest point (some 5810 vertical feet of granite), in the west; and the Dankret Range in the north. Along the western Lake Plain there are some 575,000 acres of water during the dry season, and some 3 million acres of water during the flood season. Most of the rubber plantations are located in the Lower Mekong Basin in the north and northeast of Cambodia.

The lowlands receive from five to six-and-a-half feet of rain per year. These plains are sheltered by the Cardamom Mountains, which save them from the sixteen feet of rain per year that deluges the maritime zone. The 2,500 hours of annual sunshine are important to the plain's survival, and the annual flooding of the Mekong is responsible for the two most unique resources in Cambodia: the rich soil that supports crops of rice as well as a wide variety of vegetables, and an abundance of fresh water fish.

In 1970, Cambodia produced 50,000 tons of rubber. However, the onset of civil war virtually shut down the rubber plantations, and it wasn't until the early 1990s that the plantations were reclaimed. Since then, rubber has once again become a main export.